New Products and Applications in Surfactant Technology

Annual Surfactants Review

A series of annual volumes providing in-depth reviews of the most topical areas of surfactant science and technology. Written by authors from leading laboratories around the world, it is directed at surfactant researchers and manufacturers and users of surfactants: in particular, surfactant chemists; analytical chemists; environmental chemists; users of surfactant formulations in the fields of speciality chemicals, polymers and detergents; and health and safety personnel.

Titles in the Series:

1. New Products and Applications in Surfactant Technology
Edited by D.R. Karsa.

New Products and Applications in Surfactant Technology

Edited by

DAVID R. KARSA
Akcros Chemicals
Manchester

Sheffield
Academic Press

CRC Press

First published 1998

Published by
Sheffield Academic Press Ltd
Mansion House, 19 Kingfield Road
Sheffield S11 9AS, England

ISBN 1-85075-872-7
ISSN 1460-5711

Published in the U.S.A. and Canada (only) by
CRC Press LLC
2000 Corporate Blvd., N.W.
Boca Raton, FL 33431, U.S.A.
Orders from the U.S.A. and Canada (only) to CRC Press LLC

U.S.A. and Canada only:
ISBN 0-8493-9738-3
ISSN 1097-6973

Printed on acid-free paper in Great Britain by
Bookcraft Ltd, Midsomer Norton, Bath

British Library Cataloguing-in-Publication Data:
A catalogue record for this book is available from the British Library

Library of Congress Cataloging-in-Publication Data:
A catalog record is available from the Library of Congress

Preface

Surfactants represent one of the most diverse classes of performance chemical available, being derived from a broad spectrum of organic chemicals of varying ionic type, with each product exhibiting one or more surface-active properties. Although the area of surface chemistry is a long-established technology with many well-defined products, the capacity for new products and new applications should not be under-estimated. Most of the innovative developments tend to be in industrial applications, often over-shadowed by the higher volume household detergent, consumer product and personal care sectors, which represent approximately 60% of the total surfactant market. This is understandable, since any new component of a cleaning product—especially for domestic use—must now pass a rigorous, eco-toxicological appraisal prior to use. Industrial surfactants also are now required to be more environmentally acceptable, particularly where residual surfactants may find their way into aquatic or land environments. This has led to continuous, critical assessment of the use of long-established surfactants and a search for more cost-effective, environment-friendly alternatives. Both surfactant producers and end-users have taken a voluntary, pro-active approach to such changes in recent years.

Surfactant technology represents a vibrant and challenging area of physico-chemical science, where the exact mechanisms of well-established processes are often not fully understood or are at least the subject of continued debate. This is partly due to the complex nature of even the most simple commercially available surfactants—there is still a degree of empiricism when selecting surfactants for certain applications. Journals are available which report briefly over a range of advances in this area, but there appears to be a need for this *Annual Surfactant Review* series, where important recent developments in surfactant technology are presented in a clear, informative and in-depth style, to ensure that all those interested in surface chemistry are kept fully appraised of the latest developments. With the assistance of an internationally based Editorial Board, it has been possible to commission a series of chapters from specialists around the world in their particular fields of study.

Volume I of our *Annual Surfactants Review* series includes detailed chapters on cationic, polymeric, 'hybrid' fluorocarbon/hydrocarbon surfactants and bio-surfactants, as well as on esters of alpha-sulphonated fatty acids. The production, physico-chemical properties and applications of these products are discussed in depth. Other chapters focus on specific applications, including a Japanese perspective on surfactants for personal-care products and an account of the use of surfactant systems for the environmental remediation of contaminated land. The biodegradability, aquatic toxicity and oestrogenic activity of alkylphenol ethoxylates continue to be a focus of health, safety and environmental concern and a status report is therefore an essential part of this review. Greater understanding of the physical behaviour of surfactants in solution is also an important element of the volume. Several chapters focus on this aspect of particular products and a separate chapter is devoted to the subject of multiple emulsions.

The contents of Volume I can represent only a cross-section of the major developments in this multi-chemical, multi-application area of surfactants technology. Future volumes will continue to add to and broaden the scope of our knowledge of surface chemistry.

The Editorial Board would welcome comments on both the content and format of Volume I and will be pleased to consider offers of written contributions on subjects of current interest.

D. R. Karsa

Contributors

Professor Masahiko Abe
Department of Industrial Chemistry, Faculty of Science and Technology, Science University of Tokyo, 2641 Yamazaki, Noda, Chiba 278, Japan

Professor Tomiko Fujii
Faculty of Human Life Science, Osaka City University, 3-138 Sugimoto, 3-Chome Sumiyoshiku, Osaka 558, Japan

Dr James Gadberry
Akzo Nobel Chemicals, Inc., Livingstone Avenue, Dobbs Ferry, NY 10522, USA

Professor J.L. Grossiord
Laboratoire de Physique Pharmaceutique, URA CNRS 1218, UFR des Sciences Pharmaceutiques, 5, rue Jean-Baptiste Clément, 92296 Châtenay-Malabry, France

Dr Claus-Dierk Hager
Contensio Chemicals GmbH, Hüls Group, Paul-Baumann Strasse 1, D-45764 Marl, Germany

Professor J.H. Harwell
School of Chemical Engineering and Materials Science, The University of Oklahoma, Norman, Oklahoma 73019-0628, USA

Professor R.C. Knox
School of Chemical Engineering and Materials Science, The University of Oklahoma, Norman, Oklahoma 73019-0628, USA

Dr D.A. Sabatini
School of Civil Engineering and Environmental Science, The University of Oklahoma, Norman, Oklahoma 73019-0628, USA

Mr Teruhisa Satsuki
Lion Corporation, 13-12,Hirai 7-Chome, Edogawa-Ku, Tokyo 1320035, Japan

Professor Monique Seiller
Laboratoire de Pharmacie Galénique et Biopharmacie, URA CNRS 1218, UFR des Sciences Pharmaceutiques, 5, rue Jean-Baptiste Clément, 92296 Châtenay-Malabry, France

Dr A. Silva Cunha	Faculdade de Farmacia, Universidade Federal de Minas Gerais, Avenida Olegario Macial 2360, 30180-112 Belo Horizonte, Brazil
Dr Dale S. Steichen	Akzo Nobel Surface Chemistry AB S-444 85 Stenungsund, Sweden
Dr Th F Tadros	89 Nashgrove Lane, Wokingham, Berkshire, RG40 4HE, UK
Mr Kazuyuki Yahagi	Personal Care Laboratories, Kao Corporation, 2-1-3 Bunka, Sumidaku, Tokyo 131, Japan
Professor Norio Yoshino	Department of Industrial Chemistry, Faculty of Engineering, Science University of Tokyo, 1-3 Kagurazaka, Shinjuku, Tokyo 162, Japan

Contents

1 Alkylphenol ethoxylates: biodegradability, aquatic toxicity and environmental fate

Claus-Dierk Hager

1.1 Background

Alkylphenol ethoxylates (APEs) form one of the two largest classes of non-ionic surfactants, the other being fatty alcohol ethoxylates. APEs are widely used in many institutional and industrial cleaning products as well as in industrial processes. They have been popular for their effectiveness, economy and ease of handling and formulating for more than 40 years.

APEs include nonyl-, octyl- and dodecylphenol ethoxylates, of which nonylphenol ethoxylates (NPEs) are by far the most important group commercially. APEs derived from nonylphenol comprise about 90–95% of the total world market. The world production of APEs has been estimated to be 655 000 tons per year in 1995. Table 1.1 presents an overview of the annual world demand for APEs.

Table 1.1 Annual world demand of APEs (based on 1995)

Country	APE demand (tons)
North America	210 000
Asia	200 000
South America	100 000
Western Europe	75 000
Africa	70 000
Total	655 000

Source: Hüls market research.

According to a market analysis organized by the Comité Européen des Agents de Surface et leurs Intermédiaires Organiques (CESIO) in 1995 (in which more than 20 companies were involved) the production of NPEs was 109 800 tons in Western Europe in 1994, of which 35 400 tons were exported to non-Western European countries. Thus the consumption of NPEs in Western Europe in 1994 was about 74 800 tons.

NPEs are manufactured by the base-catalyzed reaction of ethylene oxide with nonylphenol (NP) [1]. NP is produced from phenol and nonene using acid catalysis and consists almost entirely (~90%) of the para isomer. Nonene is a branched C_9 olefin manufactured by trimerizing propylene and it consists of many isomers [2]. The generic NPE structure is depicted in Figure 1.1. The average number of moles of ethylene oxide per mole of NP (n) typically ranges from 1 to 100. The most commonly used NPEs have $n = 6$–10. The NPEs manufactured to contain a specific length of the ethoxylate chain will contain a range of homologues, which follow a Poisson distribution with the peak at n. The value 'n', therefore, represents an average chain length. NPEs are slightly yellow to almost colourless liquids for the lower adducts ($n < 15$) and pale yellow to off-white pastes or waxes for higher adducts ($n > 15$). NPEs with a chain length of up to six carbon atoms are sparingly soluble in water but more soluble in organic solvents like kerosene or glycol ethers.

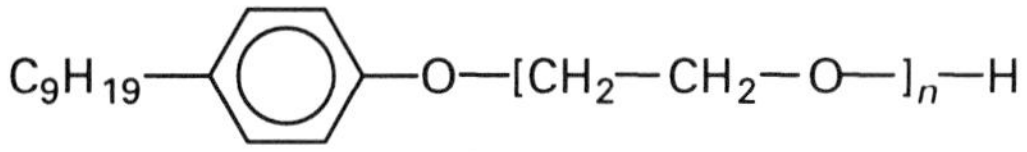

Figure 1.1 NPE structure, where n is the average number of moles of ethylene oxide per mole of NP and ranges from 1 to 100.

The major NPE markets of Western Europe are shown in Table 1.2. The industrial and institutional cleaning market consists mainly of commercial laundry products, hard-surface cleaners, metal cleaners and commercial vehicle cleaners. Since the water-soluble NPEs with 6–10 moles of ethylene oxide per mole are the most commonly used products, they are the most likely to enter the aquatic environment. It has been estimated by the Department of the Environment that in the UK 37% of the total production of these NPEs may eventually enter the wastewater system. The remainder is either destroyed or enters land disposal. In emulsion polymerization it is mainly NPEs with 20–40 moles of ethylene oxide that are used in acrylic and some vinyl acetate polymerization processes. NPEs act as a stabilizer for the final latex. In the textile and leather industries NPEs serve a variety of functions, e.g. the processes of cleaning, spinning, weaving and wetting. In the agricultural sector NPEs are used as emulsifiers, activity enhancers and wetting agents to improve the safe application of pesticides.

Table 1.2 NPE usage by industrial sector in Western Europe in 1994

Sector	NPE usage (tons)	Percentage of total usage
Industrial and institutional cleaning	26 500	35%
Emulsion polymerization and paints	10 900	15%
Textile auxiliaries	8 900	12%
Leather auxiliaries	7 200	10%
Agrochemicals	5 700	7%
Miscellaneous[a]	15 600	21%
Total	74 800	100%

[a] For example, the concrete industry, oil field chemicals, moulding agents, the paper industry.
Source: CESIO.

Because of the branching of the nonylphenol group and the presence of the aromatic ring, NPEs have been a controversial subject of discussion for many years. Numerous laboratory biodegradation studies have indicated a low biodegradability for NPEs, while others have shown high biodegradability [3]. NP and low NPE concentrations have been found in wastewater treatment plant sludges and downstream sediments in Switzerland [4]. Degradation intermediates identified as short-chain NPEs and NP carboxylate metabolites, as well as NP, were found to be much more toxic to aquatic organisms than the original NPE [5]. All of these adverse reports have created doubts about the environmental safety of NPEs. In Western Europe calls for restrictions and bans on their use started in the mid 1980s. Detergent producers in Germany and Switzerland have declared that they are willing to voluntarily stop using APEs for certain applications because of their short-chain, more toxic degradation products. In 1992 the Paris Convention (PARCOM) encouraged contracting states to prohibit voluntarily the use of APEs in domestic cleaning applications by 1995 and in industrial and institutional cleaning applications by 2000. In the USA there are no regulations relating to the use of APEs in any of these applications.

1.2 Biodegradability

The biodegradability of APEs is the major issue in terms of their possible impact on the environment. The primary biodegradability (the disappearance of the parent molecule as monitored by a specific analytical method) has been measured in numerous reported studies, both in the laboratory and in monitoring trials carried out at sewage treatment works. From these studies it is evident that both in the laboratory and in actual sewage treatment practice APEs show a high level of primary biodegradation and, specifically, they meet the biodegradability requirements of EEC Directive 82/242/EEC on the biodegradability of non-ionic detergent surfactants (copies of this Directive are available from the OECD, Brussels). Because of the variety of test systems and conditions, the data obtained are diverse and are not comparable from study to study. Some studies that illustrate the effect of various parameters on biodegradation rates are summarized in Table 1.3 and are briefly described here.

Laboratory and field studies have both shown that the primary biodegradability of APEs may drop under conditions of stress and, in particular, in trickling filter sewage works at winter temperatures levels of NPE in the final effluent may be sufficient to cause foam in receiving waters [16]. Stiff *et al.* [17] showed that at a test concentration of 20 mg/l the biodegradability of octylphenol with 8 moles of ethylene oxide per mole (OPE_8), as measured in a laboratory-scale activated sludge unit operated on a detergent-free settled sewage, fell from 95% at 15°C to approximately 70% at 11°C and to 50% at 8°C. However, at a test concentration of 5 mg/l the biodegradability of the OPE_8 was not significantly affected by temperature. Mann and Reid monitored the biodegradation of an octylphenol ethoxylate at a small trickling filter plant in the UK [7]. These trials showed that a reduction in the APE degradation from >80% to 20–25% could be correlated with a decrease in the operating temperature of the sewage treatment plant from 15°C to 5°C.

Table 1.3 Primary and ultimate biodegradation (laboratory studies)

Chemical	Test system	Extent of biodegradation	Reference
OPE_{10}	River-water die-away	90–95% in 4–24 days	[6]
	Continuous-flow activated sludge	95–100% in 2–3 days	[6]
OPE_9	Continuous-flow activated sludge	80–90% in 3 h	[7]
NPE_9	Continuous-flow activated sludge, BIAS	30–40% in 3 h	[8]
NPE_{10}	Inoculated medium, BIAS	90% in 8 days	[9]
		30% in 8 days	
NPE_{30}	Inoculated medium, BIAS	31% in 8 days	
		25% in 8 days	
NPE_4	Semi-continuous activated sludge, DOC	58% in 24 h (20 mg C/l)	[10]
NPE_9	Semi-continuous activated sludge, DOC	55% in 24 h (20 mg C/l)	
NPE_{10}	Semi-continuous activated sludge, DOC	68% in 24 h (20 mg C/l)	
NPE_{11}	Inoculated medium, DOC	80% in 30 days	[9,11]
NPE_{23}	Inoculated medium, DOC	74% in 30 days	
NPE_9	River die-away test	96–99% in 10 days	[12]
NPE_{12}	Closed-bottle test (O_2)	30% in 28 days	[13]
	Sturm (sludge) (CO_2)	65% in 28 days	
	Gledhill (sludge) (CO_2)	45% in 28 days	
	Gledhill (accl. sludge) (CO_2)	42% in 28 days	
NPE_9	Continuous-flow activated sludge (CO_2)	14–34% (10–50mg/l)	[14]
^{14}C-NPE_9	River die-away test	Half-life 42–73 days	[15]

BIAS = bismuth iodide active substances.
C/l = carbon per litre.

Lashen and Booman reported that OPE_{10} was degraded in excess of 90% in an activated sludge sewage treatment plant and that this degradation was not adversely affected by an increased plant loading and decreased detention time [6]. They also showed the effect of acclimatization in both laboratory-scale activated sludge tests and in river water samples, and in the latter context concluded that rivers that regularly receive APE surfactants contain acclimatized microflora, which rapidly biodegrade the APEs. The acclimatization of microflora depends on the test method employed and the source of river water or sludge. Saeger *et al.* point out that under the same laboratory conditions tests carried out at different times have variable results [10]. In a comparison of three separate studies with NPE_9, dissolved organic carbon (DOC) removals were 20–50%, 25–45% and 58%.

The primary biodegradation of APEs is not only influenced by temperature but also by the sludge retention time (SRT). Birch demonstrated that the biodegradation of NPE_{10} was extensive at temperatures of 15 and 11°C but was significantly reduced at 7°C [18, 19]. The results suggested a critical SRT of 6 days at low temperatures as at 2 and 4 days high levels of non-ionics were observed.

In contrast to the primary biodegradation of APEs, which is typically >90% complete under varying conditions, the ultimate biodegradation is a slower and more complex process. Hughes *et al.* compared the biodegradation of NPE_{12} in three standard test systems [13]. The ultimate biodegradation, expressed as CO_2 evolution, varied between 42 and 65% depending on the chosen system. In general, the highest biodegradation potential was achieved in a continuous-flow activated sludge unit, however, Hughes *et al.* also reported biodegradation rates of between 14 and 82%, with an average of 30 to 60% [13]. This is considered the 'typical' range of ultimate biodegradation of NPEs using the pseudo-Organisation for Economic Co-operation and Development (OECD) test methodology [20]. Kravetz *et al.* found an ultimate biodegradation for NPE_{9-10} of only 14% by measuring the CO_2 evolution in a single 28-day test [14], while Brüschweiler and co-workers obtained values of 80 and 74% by measuring the DOC of NPE_{11} and NPE_{23}, respectively [9, 11]. These workers also found a ring degradation, measured by UV absorption, which was approximately 63% complete within 30 days.

Ekelund *et al.* studied the ultimate degradation of NP in seawater at 11°C using ^{14}C-labelled phenol carbons [21]. Without sediment the NP was initially degraded very slowly, only 0.06% degraded per day, but after 28 days the rate increased to 1% per day, presumably because of adaptation of the microbial culture. When sediment was present, 1.2% of the NP degraded per day, with no increase in the rate with longer incubation times. Approximately 45% of the NP was released as $^{14}CO_2$ after 58 days, resulting in a half-life of 50 days. This experiment indicates that NP is degraded in seawater, although the breakdown is slow. Similar data were obtained by Lee *et al.*, who investigated the degradability of NP and found a 62% loss in 28 days and a half-life of 20 days [22].

Further biodegradation studies on NPEs have been carried out by Varineau and Williams using ^{14}C-ring radiolabelled NPEs [15]. The river die-away and semi-continuous activated sludge (SCAS) studies used were never intended to follow standard protocols. However, the study design was sufficient to demonstrate cleavage of the NPE phenol ring in a river environment relevant to real-world conditions. This study supports the idea that NP should not be considered 'persistent'. Although the study was not intended to derive biodegradation kinetics, the river die-away experiment showed two degradation phases: evolution of 50% $^{14}CO_2$ within 60 days followed by a slower degradation phase between 60 and 128 days during which an additional 10% of the ^{14}C NPEs was converted to $^{14}CO_2$. There are at least two possible explanations for the two-phase degradation kinetics: (i) the conversion of NPE to two distinct ^{14}C species with different half-lives of mineralization and, more likely, (ii) the loss of biomass viability in the laboratory river die-away system that was not replenished with bacterial inoculum or nutrients. Data in the river die-away experiment showed that the test system with a degradation half-life of approximately 70 days retained only 25% of its initial bacteria colony-forming (BCF) potential after 128 days, whereas the system associated with a half-life of 40 days retained 88% of its initial BCF potential. Thus, the reported half-lives for the NPE biodegradation are between 40 and 70 days.

1.3 Pathways of biodegradation

The major pathway of primary biodegradation for APEs is the stepwise oxidation of the ethylene oxide chain, either by hydrolysis or an oxidative-hydrolytic mechanism. The ethoxylate groups are progressively removed, either by ether cleavage or by terminal alcohol oxidation followed by cleavage of the resulting carboxylic acid. This results in the presence of alkylphenol, alkylphenol mono- and diethoxylate (APE_1 and APE_2) and carboxylate metabolites ($APE_{1-2}C$ and APC), which have been detected in laboratory studies and in treated wastewater effluents and sludge [20]. Based on the fate of APE in studies by Yoshimura [12], Ball *et al.* [23] and Varineau and Williams [15], and particularly the large number of field studies by Marcomini and Giger [4], the biodegradation pathways of APEs are shown in Figure 1.2.

In most studies the short-chain intermediates, and particularly the corresponding carboxylic acids, are regarded as being of low biodegradability or even persistent [3, 12]. Williams *et al.* studied the biodegradation of the ether carboxylate degradation intermediates of nonylphenol and octylphenol ethoxylates using the standard OECD method 301 B (modified Sturm Test, observing CO_2 evolution) [24]. These were the first biodegradation studies performed on an ether carboxylate degradation intermediate of either nonylphenol or octylphenol ethoxylates alone. The test was carried out using mainly non-acclimatized seed. For both octylphenol ether carboxylates (OPEC) and nonylphenol ether carboxylates (NPEC), more than 60% of the theoretical carbon dioxide amount was evolved within 28 days (Table 1.4). These results are surprising in the light of the common perception that ether carboxylates are persistent intermediates in the degradation of APEs but they are consistent with field studies showing extremely low levels of NPEC in rivers exposed to NPE [25, 26].

The ultimate biodegradation of the branched alkyl chain moiety is not described in any studies. Studies using radiolabelled ^{14}C atoms are limited to the phenol moiety or ethylene oxide units. However, Swisher reports that branched fatty acid chains can be oxidized by a combination of β- and α-oxidation [3].

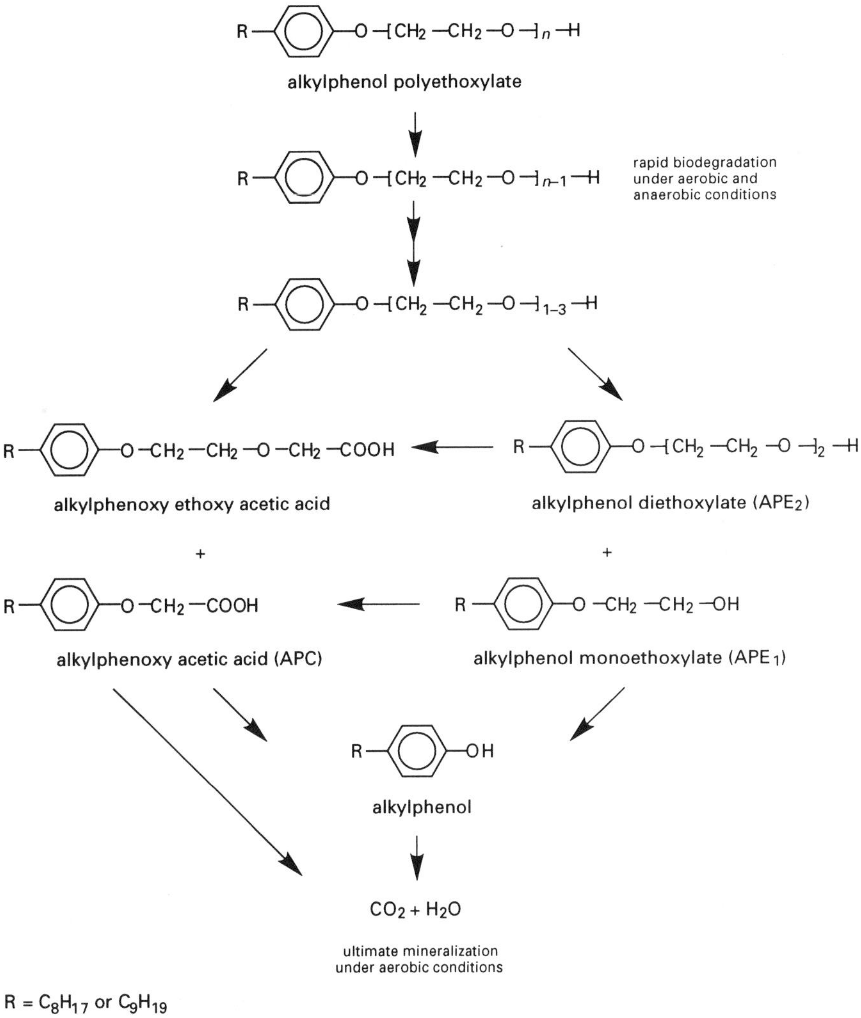

Figure 1.2 Pathways of biodegradation of APEs.

Table 1.4 Biodegradation of OPEC and NPEC according to OECD method 301 B (CO_2 evolution) for non-acclimatized seed

Product	Extent of biodegradation	Half-life
OPE_1C	72% loss in 28 days	15 days
OPE_2C	80% loss in 28 days	12 days
NPE_1C	59% loss in 28 days	22 days
NPE_2C	65% loss in 28 days	18 days

Source: Williams *et al.* [24]

1.4 Removal of NPEs in sewage treatment plants

In most countries of Western Europe, as well as in the US, waste water generally receives biological treatment. The proportion of waste water receiving sewage treatment in Western Europe is approximately 70%, although it exceeds 80% in Germany, Switzerland, The Netherlands and Sweden. APEs are biodegraded during aerobic and anaerobic treatment of waste water to alkyl phenols, alkylphenol mono- and diethoxylates, the corresponding carboxylates, CO_2 and water. The reported results vary widely from poor treatability in some Swiss treatment plants [27] to over 95% removal [28]. A detailed overview is given by Talmage [20].

Ahel *et al.* made a detailed study of the fate of NPEs and their metabolites in 11 sewage works in Switzerland (the collection of samples was made in 1983–86) [29]. Table 1.5 presents the mass flows (in mol/day as well as in kg/day) of nonylphenolic compounds for the entire sewage treatment process at the sewage treatment plants (STPs) Uster and Zurich-Glatt. The raw, untreated sewage was taken as input, with secondary effluent and digested sludge as output. Ahel *et al.* found that approximately 60–65% of all nonylphenolic compounds (calculated on molar basis) introduced to STPs are discharged into the environment [29]. Furthermore, approximately 60% of these discharged compounds are lipophilic metabolites (NP, NPE_1 and NPE_2), which are well known for their enhanced toxicity to aquatic organisms. The distribution of nonylphenolic compounds in the digested sludge was 95% NP and 5% NPE_1 + NPE_2; the low percentage of NPE_1 and NPE_2 is the result of their hydrophobic nature. The more hydrophilic compounds were present in the secondary effluent of the Zurich-Glatt plant, with the nonylphenolic compounds distributed as follows (on a molar basis): 48% NPE_1C, NPE_2C, 26% NPE_1, NPE_2, 20% $NPE_{n>2}$ and 8% NP [29]. The highest elimination rates were achieved in the STPs characterized by low sludge-loading rates and nitrifying conditions.

Mann and Reid found that the elimination of NPE under real conditions can be much higher in summer than in winter because of the dependency of biodegradation on temperature [7]. Data from Ahel *et al.* confirm these observations and are shown in Table 1.6 [29]. In winter NPE_2 elimination was lower than its formation.

Many different plant designs are used but most municipal waste water is treated by some form of activated sludge system. Consequently, understanding the treatability of NPE in these plants is a key aspect of understanding the environmental fate of these surfactants. Naylor *et al.* studied the biological treatability of NPEs in seven different wastewater treatment plants in the US, including a pair of STPs in the southeast US, one of which received discharge mainly from textile mills and furniture factories and the second received mainly

Table 1.5 Mass flows of NPEs and their metabolites in sewage treatment plants (identical studies expressed in the top table in mol/day and in the lower table in kg/day)

	NPE + metabolites (mol/day)				Released to environment	
STP	Influent	Effluent	Sludge	Biodegradation	Effluent	Sludge
Uster	30.2 (100%)	11.6 (38%)	6.4 (21%)	41% loss	64%	36%
Zurich-Glatt	94.5 (100%)	39.6 (42%)	218 (23%)	35% loss	65%	35%

	NPE + metabolites (kg/day)				Released to environment	
STP	Influent	Effluent	Sludge	Biodegradation	Effluent	Sludge
Uster	14.2 (100%)	3.7 (26%)	1.4 (10%)	64% loss	72%	28%
Zurich-Glatt	54.0 (100%)	11.7 (22%)	4.8 (9%)	69% loss	71%	29%

Source: Ahel *et al.* [29]

Table 1.6 Temperature dependence of NP, NPE_1 and NPE_2 elimination efficiency in sewage treatment plant Zurich-Glatt

Date	Temperature (°C)	Elimination (%)		
		NP	NPE_1	NPE_2
31/01/1984	10–13	65	18	−2
20/06/1983	18	75	40	25
27/08/1984	20	82	72	47

[a] Date of sampling, 24 h composite sample.
Source: Ahel *et al.* [29].

domestic sewage [28]. The study also included an STP in a midwestern city that received discharge from a cleaning product manufacturer as well as domestic sewage and was sampled in both summer and winter. All the STPs used activated sludge digestion. Loadings of NPE ranged from approximately 1100–2400 μg/l in the municipal plants up to 33 700 μg/l in the paper-mill treatment plant. The NPE removal in the STPs was consistently well above 90% (range 92.5–99.8%) with no accumulation of NP or the lower NPEs in the effluent or sludge. The overall NP concentration in the effluent ranged from <0.20 to 15.3 μg/l and for $NPE_{1\text{-}17}$ from < 5 to 261 μg/l. Naylor *et al.* pointed out that oligomer distribution of NPE in influents and effluents can be a sensitive measure of degradation conditions [28]. The more hydrophilic species (the intermediate NPE oligomers and the carboxylic acids) can be expected to remain primarily in the overlying water, but NP, NPE_1 and NPE_2 have a greater tendency to adsorb on the sediments. The higher concentration of oxygen present in river waters may also tend to produce a larger fraction of carboxylic acid derivatives in the water, while the generally lower oxygen concentrations in bottom sediments may favour formation of NP along with the short-chain ethoxylates, with a relatively small amount of carboxylic acids [30]. The removal rates in the sewage plants were not diminished by higher loadings. In addition, the seasonal effects were minor. The two winter samples had removal rates 2–5% less than those of corresponding summer samples [31]. A summary of the results is shown in Table 1.7.

Blackburn and Waldcock studied the total NP concentrations of eight rivers and 16 sewage effluents in the UK and found them to be in the range <0.2 to 12 µg/l [32]. Concentrations of dissolved NP were in the range <0.10 to 5.4 µg/l. Similar data have been found by Lee and Peart for four Canadian municipal treatment plants [22]. The concentrations of NP in the effluent ranged between 0.80 and 15.1 µg/l. Clark *et al.* analysed the effluent of three US municipal treatment plants in New Jersey [33]. The concentrations of NP in the effluent were in the range 1.0–5.0 µg/l, for NPE_2 the range was 1–123 µg/l and for NPE_{3-5} 1–25 µg/l. In Italy Crescenzi *et al.* investigated the removal of NPEs in three STPs in Rome during winter and summer [34]. The average removal of all NPEs in February was 96.9% compared to 98.6% in June. These data are consistent those of Naylor [31]. The overall effluent concentration for NPEs was 0.74–12.0 µg/l.

Varineau and Williams [15] investigated the biodegradation of ^{14}C-ring radio-labelled NPE_9 in a SCAS system to evaluate whether cleavage of the phenol ring occurs during NPE biodegradation in an activated sludge environment. The SRT of 52 days was longer than expected from standard protocols but did approximate reasonably to that of an activated sludge treatment environment. The mass balance is shown in Table 1.8. Overall, a 93–97% removal of NPE was obtained from the SCAS unit. A significant portion of the ^{14}C in the effluent consisted of soluble metabolites that had degraded beyond the phenol ring. This result confirms several studies on wastewater treatment plants that show efficient removal of NPE [20, 31] as well as demonstrating ring cleavage under activated sludge conditions that are relevant to real-world conditions.

Table 1.7 Wastewater treatment plant treatability of NPEs

Plant and location	Date	Removal rate of influent/effluent (wt %)	Influent NP (μg/l)	Influent NPE_{1-17} (μg/l)	Effluent NP (μg/l)	Effluent NPE_{1-17} (μg/l)
Burglington, NC						
East Plant	May 1992	94.6	384	903	14.4	54
South Plant	May 1992	98.2	359	2619	15.3	39
High Point, NC[a]						
East Plant	Jan. 1993	96.5	44	289	< 1	< 5
West Plant	Jan. 1993	97.7	124	487	1	< 5
East Plant	April 1993	98.3	14	273	< 0.2	< 5
West Plant	April 1993	99.2	12	631	< 0.2	< 5
East Plant	July 1993	99.3	234	640	< 1	< 5
West Plant	July 1993	99.8	978	3140	0.2	< 5
East Plant	May 1988	95.6		1780	1.3	79
West Plant	May 1988	95.6		2395	2.3	95
Small city, midwest USA[b]	August 1990	97.2	121	1390	3.5	39
Small city, midwest USA[b]	March 1991	92.5	28	1110	4.9	80
Wood pulp mill, USA	June 1990	97.6	24	8450	3.3	201
Paper mill, USA	April 1993	99.3	516	33700	8.4	261

[a] Received discharge from textile and furniture industries as well as domestic sewage.
[b] Received discharge from a cleaning product manufacturer as well as domestic sewage.
Source: Naylor [28].

Table 1.8 Biodegradation of ^{14}C-ring radiolabelled NPE in an SCAS mass balance

Input	Output				
^{14}C NPE	$^{14}CO_2$	Sludge removal	Effluent	Remaining in the STP	Lost
100% (2 mg/l)	20.8% in 30 days	0.72% ^{14}C	6.9% ^{14}C NPE 6.9% methylene chloride extractable (non-NPE) 19.1% acid extractable 23.1% non-extractable (short-chain, polar compounds)	14.35% ^{14}C in system	8%
100%			100%		

Source: Varineau and Williams [15].

1.5 Fate and levels in sludge

NP and, to a somewhat lesser extent, NPE_1 and NPE_2 are more hydrophobic than common NPEs, as indicated by their measured octanol/water partition coefficient (log K_{ow}) values of 4.5, 4.2 and 4.2, respectively [35, 36]. As they are less water soluble than the higher molecular weight oligomers, they are partially removed from the wastewater stream by adsorption to lipophilic flocs of the sewage sludge. Thus, concentrations of these short-chain oligomers, as well as NP, are usually higher in sludge than in waste water. Giger *et al.* [37] and Brunner *et al.* [38] showed that the concentrations of NP in anaerobically stabilized sludge were higher than in aerobically stabilized sludge, averaging 1010 mg/kg (based on dry weight) compared to 280 mg/kg for aerobically stabilized sludge. They concluded that NP was not significantly degraded under anaerobic conditions; on the contrary, it was accumulated in the digested sludge to concentrations exceeding 2000 mg/kg. Compared with the input load, the NP enrichment factor in anaerobically digested sludge was 4.7–6.4 [39]. However, studies in the US have shown that treatment of NPE in waste water can be quite efficient, with little residue remaining in the sludge. These data indicate much lower levels (10 mg/kg dry weight) than those reported by Ahel *et al.* [29, 40] and Naylor *et al.* [28, 31] and hence greater removal rates. Complete biodegradation of the short-chain ethoxylates and their derivatives appears possible under aerobic conditions [29]. Under anaerobic conditions Ejlertsson *et al.* studied the behaviour of ^{14}C-ring radiolabelled short-chain NPE in sludge [41]. NPE_1 and NPE_2 were transformed to NP but the NP formed was not thought to degrade further during the experimental period (20 weeks) since no $^{14}CO_2$ or $^{14}CH_4$ could be observed. Table 1.9 shows a summary of NP, NPE_1 and NPE_2 concentrations in sludge.

Table 1.9 Summary of NP, NPE_1 and NPE_2 concentrations in sludge

Country	NP (mg/kg dry weight)	NPE_1 (mg/kg dry weight)	NPE_2 (mg/kg dry weight)	Source
Switzerland	450–2530 (anaerobe)	—	—	[37]
	80–500 (aerobe)	—	—	[37]
Switzerland	460–1200	180–230	30–40	[4, 42]
Switzerland	1270 (anaerobe)	183 (anaerobe)	44 (anaerobe)	[38]
	304 (aerobe)	362 (aerobe)	132 (aerobe)	[38]
USA	10	—	—	[28]
Germany	4.6–214	—	—	
Germany	90–1300	—	—	[43]
Switzerland	200	—	—	[44]

1.6 Fate in soil

Sewage sludge is often disposed of in landfill sites or used as a fertilizer on land, after anaerobic digestion. Since the sewage sludge can contain rather high levels of NP and short-chain ethoxylates, it is important to establish the fate of these compounds in soil. The degradation rates of chemicals in soil will vary with soil type, temperature, nutrients and other factors. Trocmé *et al.* described one test in which NP in soils decreased 89% in 40 days (100 mg/kg initial dose) and a second test in which the decrease of NP was 62% in 40 days (1000 mg/kg initial dose) [45]. Marcomini *et al.* examined the fate of NP, NPE_1 and NPE_2 in soil which had been treated with sludge [42]. The initial concentrations in the soil were 4.7, 1.1 and 0.1 mg/kg (dry weight), respectively. These initial concentrations decreased by about 90% after one year (to 0.5, 0.12 and 0.01 mg/kg); however, 80% of the reduction occurred during the first month. In spite of the reduction of the contaminants in the sludge-amended soil, the authors noted that there was still some persistence. The authors hypothesize that biodegradation occurs in the aqueous phase and on the surface of the soil particles, and eventually contaminant residues become incorporated in the soil particles or are strongly bound to the organic matter, making them less bioavailable.

The Fraunhofer Institute has carried out a study undertaken to follow the potential for mobilization and degradation of NP, NPE_1 and NPE_2 in soils treated with sludge. The soils chosen were from Brandenburg (Germany) and were classified as weakly sorbent sandy soils. The soils were treated with two sludges, one low in surfactants and the other containing an average concentration of surfactants. The results of the study showed that there was no recognizable washout of NP or NPE_1 and NPE_2, all were fixed in the top 10 cm layer of the soil. It was also shown that the NPEs were rapidly biodegraded, initially to NP, and that within 20 days there was no residue detectable. The NP data were similar, except that there was a maximum in the concentration after 7 days, probably as a result of the breakdown of NPE. However, within 20 days there was no longer any detectable NP in the soil (Penman, unpublished work). Similar results in terms of biodegradation have been reported by Hughes *et al.*, who studied the biodegradation of NPE in soil [46]. The results of this study suggest that NPE biodegrades ultimately and completely (ring cleavage) in soil.

1.7 Fate and levels in rivers and sediments

Ahel *et al.* investigated the behaviour of NPEs and their metabolites in the Glatt River (close to Zurich) in Switzerland [40]. Discharge of secondary effluents from municipal STPs into the river was the predominant source of the investigated compounds. Based on composite sampling conducted during summer (1983–86), a mass balance of nonylphenolic compounds in the Glatt River was calculated. The total input coming from secondary effluents was 108.2 mol/day (35 kg/day). The hydraulic retention time, for river water containing the majority of nonylphenolic compounds, was about 10–15 h (the retention time in an STP is about 6–8 h). The total output of nonylphenolic compounds from the Glatt River (immediately before the river mouth at Rheinfelden/Bale) was 82 mol/day. This results in an overall biodegradation efficiency of only 24% on this 35 km long river section (Table 1.10). The removal rate was significantly lower than the range of 26 to >90% elimination efficiency found for STPs. A negative elimination was observed for NPE_1C+NPE_2C, which formed from NPE. Ahel *et al.* observed significant compositional changes in the river: a decrease for NPE_{3-20} from 21 to 3.5% and an increase for NPEC from 51 to 85% (Table 1.11) [40].

Table 1.10 Mass flows of NPEs and their metabolites in the Glatt River, Switzerland

	Mass flow (mol/day)				
	NPE_{3-20}	NPE_1+NPE_2	NP	NPE_1C+NPE_2C	NP total
Input[a]	23.4	24.3	5.3	55.2	108.2
Output[b]	2.8	7.2	2.0	70.1	82
Elimination (%)	88	70	62	−27	24

[a] Secondary effluent from 10 STPs.
[b] At Glatt Rheinfelden.
Source: Ahel *et al.* [40].

Table 1.11 Compositional change of nonylphenolic compounds in the Glatt River

	Input from 10 STPs (%)	Output near river mouth (%)
NPE_{3-20}	21	3.5
NPE_1+NPE_2	23	9
NP	5	2.5
NPE_1C+NPE_2C	51	85

Source: Ahel *et al.* [40].

In the US a comprehensive monitoring study in conjunction with the Environmental Protection Agency (EPA) was carried out with the objective of characterizing the distribution of NP and NPE concentrations in 30 river locations in the US. The sampling locations were chosen with the aid of several databases maintained by the US EPA. It should be emphasized that the reaches sampled represent a subset of locations where NP and NPE were considered most likely to occur. The locations ranged from rivers with little human impact to highly impacted rivers with both industrial and municipal effluents. NP and NPE concentrations in river water were mostly (60 to 75% of the samples) below their detection limits (about 0.1 μg/l for NP, NPE_1 and NPE_2, and 1.6 μg/l for NPE_{3-17}). The highest levels found were about 1 μg/l for NP, NPE_1, and NPE_2, and 15 μg/l for NPE_{3-17}. Significantly less NPE_1 was found in sediment and it was not detected in 44% of the samples. The majority of sediment samples contained detectable amounts of NP and NPE_1, ranging up to 3000 μg/l for NP and 170 μg/l for NPE_1. Sediment interstitial water concentrations of NP were estimated to be similar to concentrations in the water column [28].

Another study in the US was carried out in 1993–95 on the Fox River, Wisconsin. The Fox River connects an inland lake (Lake Winnebago) with Green Bay (part of Lake Michigan) and receives discharges from 15 paper mills and six city wastewater plants. The influx of NPE to the river from paper mills and cities was measured in summer and winter, but seasonal effects on treatment efficiency proved to be minor. Samples from the Fox River showed low concentrations of NP, NPE_1 and NPE_2 in sediment and generally non-detectable concentrations in water. Comparison of NPE mass fluxes at the river mouth with the influxes indicated extensive degradation in warm river water but little in cold water. The concentrations of NP, NPE_1 and NPE_2 in the Fox River were comparable to those in the rivers in the 30 rivers study where there were industrial plants and secondary treatment facilities [26]. Some statistical data describing the results of several river studies are shown in Table 1.12.

Table 1.12 Summary of reported environmental exposure data

	River			Sediment			Reference
	No. of samples	Range (μg/l)	Average (μg/l)	No. of samples	Range (μg/l)	Average (μg/l)	
Glatt River, Switzerland							
NP	110	< 0.2–45	60% < 2		190–13 100		[29]
NPE_1	110	< 0.3–69	70% < 5		100–8 850		
NPE_2	110	< 0.3–30			ND–2 720		
NPE_1C	48	1–45	80% < 15				
NPE_1C	48	2–71					
30 US rivers							
NP	98	< 0.11–0.64	0.12	81	< 2.9–2 960	162	[28]
NPE_1	98	< 0.06–0.60	0.09	81	< 2.3–175	18	
NPE_2	101	< 0.07–1.2	0.1				
NPE_{3-17}	101	< 1.6–14.9	2.0				
Fox River, US							
NP (Fall, 1993)			0.2			218	[26]
NPE (Fall, 1993)			3.5			71	
NP (Winter, 1995)			0.2			410	
NPE (Winter, 1995)			5.5			8	
NP (Summer, 1995)			0.6			140	
NPE (Summer, 1995)			2.8			46	
NPEC (Summer, 1995)			13.5			—	
River Main, Germany							
NP (1990)	47	0.007–3.3	90%: 0.08	47	56–14 800	90%: 9.5	
NP (1991)	54	0.009–1.3	90%: 0.18	54	22–13 200	90%: 7.7	
NP (1995)		0.05–0.1			100–1 000		[43]

Table 1.12 (continued)

	River			Sediment			Reference
	No. of samples	Range (µg/l)	Average (µg/l)	No. of samples	Range (µg/l)	Average (µg/l)	
Italy, several rivers							
NPE		0.64–4.30	1.5				[34]
UK							
NP (rivers)	23	0.02–1.6					[32]
NP (River Aire)	6	24–53					
NP (estuaries)	10	< 0.08					

ND = not determined.

The behaviour of nonylphenolic compounds during the infiltration of river water into groundwater has been investigated by Ahel and co-workers [36, 47] and Schaffner *et al.* [48]. The elimination of the surfactant-derived alkylphenolic compounds during infiltration of river water into groundwater is highly dependent on the predominant chemical form of such compounds in the river and specific conditions in the aquifer. For NPE_1 and NPE_2 elimination seems to be efficient (>90–95%) in all seasons, while for NP it can be reduced during winter. Some exposure data are summarized in Table 1.13.

1.8 Aquatic toxicity

In contrast to mammalian data [20], in general non-ionic surfactants are toxic to aquatic organisms. While the mechanism of toxicity is unknown, the chemical structure of surfactants is predictive of their detrimental effect. The toxicities of NP and APEs with a range of ethylene oxide units were tested in several species of aquatic organism and reported by Talmage [20]. In total there exist about 157 studies of the aquatic toxicity of NP and NPEs carried out on 93 species, of which 16 studies are long term. Most of these studies are briefly described by Talmage [20]. In all cases the toxicity of APEs to aquatic organisms increases (as shown by a decrease in the LC_{50} value, i.e. the concentration that causes 50% mortality) with a decrease in the number of ethylene oxide units and an increase in the hydrophobic chain length. Therefore, the toxicity of the parent surfactants is less than that of the degradation products or intermediates for NP, NPE_1 and NPE_2 as well as for NPECs. This has been systematically investigated by Yoshimura on Japanese killifish (*Oryzias latipse*) [12]. The 48 h LC_{50} value of NPE_{16} was 110 000 μg/l, which decreased to 1400 μg/l for NP. Yoshimura also showed that the toxicity increases again in the range where the surfactants have their best wetting performance (usually the 9 mole ethylene oxide product). Limited toxicity data on the carboxylic acid degradation products NPC and NPE_1C suggest the toxicity is low, comparable to NPE_1, probably because of their higher water solubility. Yoshimura's findings are shown in Figure 1.3 [12].

The APE Panel of the Chemical Manufacturers' Association (CMA) has sponsored a comprehensive program of testing (including acute and chronic exposures) of the toxicity of NP to aquatic life. Saltwater and freshwater fish, algae and invertebrate animals are represented in the program. All tests were done using US EPA approved protocols according to good laboratory practice. Tables 1.14 and 1.15 summarize the results of these studies as well as some information from elsewhere. Fish, invertebrates and algae all vary widely, in terms of magnitude, in their sensitivity as measured by acute toxicity. NP is approximately 10 times more toxic than NPE_9. $NPE_{1.5}$, which is primarily made

Table 1.13 Summary of reported exposure data in groundwater

	NP (μg/l)	NPE_1 (μg/l)	NPE_2 (μg/l)	NPE_1C (μg/l)	NPE_2C (μg/l)
Glatt River, Switzerland[a]					
Distance from river bed (m)					
0	2.7±2.2	7.8±5.2	8.4±5.6	14.7±8.9	24.7±5.7
2.5	0.96±0.94	0.91±1.3	0.33±0.51	10.9±3.1	22.4±1.1
5	0.4±0.43	0.18±0.25	0.16±0.29	9.7±0.2	15.3±6.6
7	0.44±0.46	0.09±0.12	0.06±0.29	2.9±0.1	14.5±2.5
13	0.20±0.32	0.04±0.09	< 0.01	4.5±6.3	5.1±7.1
Sitter River, Switzerland[a]					
River water	1.8±0.52	2.4±0.34	1.8±0.71		
Groundwater	0.09±0.05	0.12±0.18	0.07±0.14		
Elimination	94.8%	96.9%	96.2%		
River Main, Germany[b]					
Groundwater	0.047–0.15				
Well water	0.027–0.065				

Sources: [a]Giger *et al.* [27]; [b]Kalbfus and Zellner [43].

up of NPE_1 and NPE_2, is significantly less toxic than NP. The species chosen by the US EPA and the CMA have been identified as very sensitive organisms. The fathead minnow, mysid shrimp and marine alga show chronic toxic effects with NP concentrations at or below 20 μg/l.

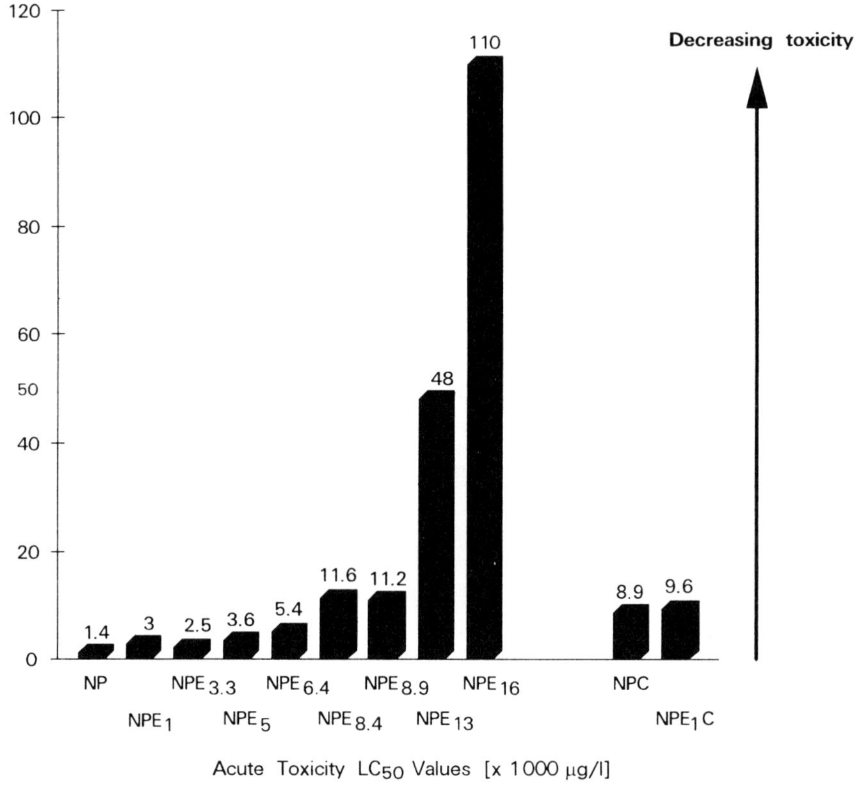

Figure 1.3 Toxicity versus ethylene oxide chain length of the NPEs. Data are 48 h LC_{50} values for the Japanese killifish (*Oryzias latipse*). From Yoshimura [12].

1.9 Bioconcentration

The bioconcentration factor (BCF) refers to the concentration of a test chemical in an aquatic organism divided by the freely dissolved water concentration. A summary of BCFs is given in Table 1.16. The BCFs (wet-weight, non-lipid basis) for fathead minnows range from 270 to 741, for salmon from 10 to 280,

Table 1.14 Comparison of aquatic toxicity of NP and NPEs

Species	Type	NP		NPE_9	
		Range (µg/l)	Typical (µg/l)	Range (µg/l)	Typical (µg/l)
Fish	Acute, 96 h LC_{50}	128–3000	200–400	1300–1 000 000	5000–9000
Fathead minnow	Acute, 96 h LC_{50}	128		4600[a]	
(*Pimephales promelas*)					
Invertebrates	Acute 24–96 h LC_{50}	20.7–3000	200–400	2900–100 000	10 000
Water flea	Acute, 48 h LC_{50}	190[b]		14 000[a]	
(*Daphnia magna*)					
Cladoceran	Acute, 96 h LC_{50}	276		$NPE_{1.5}$ 1016	
(*Ceriodaphnia dubia*)	96 h EC_{50}	69		$NPE_{1.5}$ 626	
Algae	Acute, EC_{50}	27–7500	500–1000	210–5 000 000	10 000
Green algae	Groth, 96 h EC_{50}	410[c]		12 000[a]	
(*Selanastrum capricomutum*)					

[a]Kravetz *et al.*[14], Dorn *et al.* [49].
[b]Comber *et al.* [50].
[c]CMA [51].
All other data refer to Talmage [20].

for carp from 0.9 to 3.3, and for mussels from 13 to 340 (apart from the values >1000 reported by Ekelund [57], see below). Ahel *et al.* demonstrated that algae, fish and ducks can bioaccumulate NP, NPE_1 and NPE_2 from freshwater environments [61]. Algae bioaccumulated NP to a greater degree than higher organism, possible due to adsorption. Furthermore, the BCFs calculated by Ahel [61] are given on a dry-weight basis but the usual convention is to base them on fresh weights, which are about 10 times lower, depending on the organism.

Table 1.15 Chronic aquatic toxicity of NP

Species	Effect	NOEC[a] (μg/l)	LOEC[b] (μg/l)	Reference
Fish				
Fathead minnow	Growth, 28 day	23	> 23	[52]
(*Pimephales promelas*)	Survival, 33 day	7.4	14	
Rainbow trout (*Salmo gairdneri*)	Growth and survival, 91 day	6.3	10.3	
Invertebrates				
Water flea (*Daphnia magna*)	Survival, 21 day, growth, reproduction	24	39	[50]
Marine shrimp	Growth, 28 day	3.9	6.7	[53]
(*Mysidopsis bahia*)	Survival, 29 day	6.7	9.1	
Cladoceran	Reproduction, 7 day	89	202	[54]
(*Ceriodaphnia dubia*)	Survival, 7 day	202	377	
Data for $NPE_{1.5}$	Reproduction, 7 day	285	886	
	Survival, 7 day	285	886	
Algae				
Green algae (*Selanastrum capricornutum*)	Growth (96 h EC_{50})	92	190	[55]
Marine algae (*Skelatonema costatum*)	Growth (96 h EC_{50})	10	20	[51]

[a]NOEC = no observable effect concentration.
[b]LOEC = lowest observable effect concentration.

Table 1.16 NP bioconcentration factors

Fish	BCF	Invertebrates	BCF
Bluegill sunfish	220	Mussels	13
Carp	0.9–3.3		340
Fathead minnow	270		2700 (NP only)
	350		3430 (NP+metabolites)
	741		
Salmon	10–280	Shrimp	1.4–7.9 (96 h)
Stickleback	1250		90–110
BCF predicted from: log K_{ow} = 4.48 and 5% lipid[a]	1510	BCF predicted from: log K_{ow} = 4.48 and 1.75% lipid[a]	530

[a] $BCF^{fd}_{lipid} = K_{ow}$ [56].
Sources:
Ekelund *et al.* [57]
McLeese *et al.* [58, 59]
CMA [52, 53, 60].

A bioconcentration test on NP in the fathead minnow (*Pimephales promelas*) has been carried out by the CMA [52]. The fathead minnow was exposed to either 4.9 or 22.7 µg/l of NP under flow-through conditions for 20 days and then to dilution water without test substance for 7 days. Exposure of the fathead minnow to 4.9 µg/l of NP in these conditions resulted in a BCF of 271, with an uptake rate constant of 133 and a release rate constant of 0.49. Exposure of the fathead minnow to 22.7 µg/l of NP under the same conditions resulted in a BCF of 344, with an uptake rate constant of 193 and a release rate constant of 0.56 [52, 60]. The time taken to reach 90% steady-state conditions was 4.5 and 4.1 days, respectively. The rate of uptake of NP and the rate of clearance after transferring the fish to clean water were both rapid, with half-lives of 1.2 to 1.4 days [28, 30].

The highest BCF values found in the literature were reported by Ekelund *et al.*, who calculated BCFs of 1250 for fish and 3430 for mussels [57]. However, these values were determined using radioactive carbon-labelled NP and were based on total radioactivity, which means that the BCFs reflect retention of both parent NP and all its metabolites in the body and therefore constitutes an upward bias and overestimate of the true BCFs. Additionally, the mussel data never reached equilibrium in the test so the results are only estimates of the actual BCF. The study by Ekelund *et al.* also showed that release was at least as rapid as uptake in stickleback and mussels, and was much faster than uptake in shrimps [57]. These data, coupled with similar findings of rapid release and rapid metabolism in higher animals like mammals, show that NP is not highly bioaccumulative and does not biomagnify in the food chain (Penman, unpublished work).

1.10 Estrogenic effects of NPEs

An extensive literature review on the estrogenic effects of APEs has been written by Nimrod and Benson [62]. Evidence of a weak estrogen-like action of some alkylphenols has been known for many years. Mueller and Kim have shown that several alkylphenols are capable of displacing estradiol from the estrogen receptor [63]. However, evidence from these experiments did not prove that the interaction at the estrogen receptor was one of direct competition for a single binding site. Mueller and Kim did find that the ability to displace estradiol depended on *para* substitution of the phenol and the displacement ability was increased with increasing hydrophobic chain length [62, 63].

In 1992, investigators in the UK reported that in male fish NP was capable of inducing the production of vitellogenin, a yolk protein that is normally produced in response to estrogen in female fish (Sheahan and Harries, unpublished work). Several *in vitro* assays have been developed for estrogenic potency and applied to alkylphenols with somewhat uneven results. In general, NP shows some estrogenic activity in these assays when using dose levels that are several orders of magnitude higher than the traces of NP and NPEs that have typically been reported in surface waters. For example, Jobling and Sumpter showed that at the high doses examined (11 000–44 000 μg/l NP) the alkylphenols were cytotoxic [64]. Jobling and Sumpter found that the potency of NP in the assay was more than five orders of magnitude smaller than that of estradiol, a natural estrogen [64]. The data also showed a decreasing potency with longer ethylene oxide units (Table 1.17).

The estrogen receptor binding affinity of NP and select NPE degradation products in comparison to estradiol has been measured by White *et al.* [65]. The binding affinities of NP and NPEs were 3 000 to 20 000-fold less in comparison to estradiol (the concentration of 200 μg/l used in this test far exceeded environmental concentrations). The results of other binding studies suggest that the environmental estrogens examined, including phytoestrogens in plants and vegetables, which are part of our daily diet, have a similar affinity for the receptor, about 500- to 5000-fold less than estradiol's receptor affinity [62].

Table 1.17 Estrogenic potency data developed by Jobling and Sumpter [64] using rainbow trout in *in vitro* hepatocyte bioassay

Compound	Potency values relative to estradiol
Estradiol	1
4-tert-butylphenol	0.00016
4-tert-octylphenol	0.000037
NP	0.0000090
NPE_1C	0.0000063
NPE_2	0.0000060
NPE_9	0.0000002

This and other *in vitro* methods may be useful as screening methods but the importance of any possible estrogenic effects of NP can only be judged by testing relevant concentrations in *in vivo* studies. Jobling *et al.* exposed male rainbow trout to NP at 30 μg/l in a flow-through system for three weeks [66]. After that time the fish were killed and testicular weight measured. The vitellogenin production was increased significantly at 20 μg/l NP, but the importance of this detectable increase in terms of the health of the fish is still uncertain. In addition, the observed inhibition of testicular growth provides an end-point that is clearly an adverse effect, with a lowest observed effect concentration (LOEC) of 54.3 μg/l, and a no observed effect concentration (NOEC) of 20.3 μg/l.

Given the low reported environmental concentrations and BCFs of NP and NPE products, the potential for these compounds to produce estrogenic effects in the environment appears low. As there are still many questions concerning the human and environmental risks involved with NP and other suspected endocrine chemicals, a world-wide research programme has been set up that should lead to a better understanding of the biological effects of environmental estrogens on human and animal life.

References

[1] Cahn, A. and Lynn, J.L. (1978) Surfactants, in *Kirk-Othmer Encyclopedia of Chemical Technology*, vol. 22, John Wiley & Sons, Inc., New York.

[2] Reed, H.W.B. (1978) Alkyl phenols, in *Kirk-Othmer Encyclopedia of Chemical Technology*, vol. 2, John Wiley & Sons, Inc., New York.

[3] Swisher, R.D. (1987) *Surfactants Biodegradation*, vol. 18, Surfactant Science Series, Marcel Dekker, Inc., New York.

[4] Marcomini, A. and Giger, W. (1987) Simultaneous determination of linear alkylbenzene sulfonates, alkylphenol polyethoxylates and nonylphenol by high-performance liquid chromatography. *Analytical Chemistry*, **59** 1709-15.

[5] Giger, W. and Stephanou, E. (1982) Persistent organic chemicals in sewage effluents. 2. Quantitative determination of nonylphenols and nonylphenol ethoxylates by glass capillary gas chromatography. *Environmental Science and Technology*, **16** (11) 800-805.

[6] Lashen, E.S. and Booman, K.A. (1967) Biodegradability and treatability of alkylphenol ethoxylates: a class of nonionic surfactants. *Water Sewage Works*, **114** R155-R163.

[7] Mann, A.H. and Reid, V.W. (1971) Biodegradation of synthetic detergents: evaluation by community trials. II: alcohol and alkylphenol ethoxylates. *Journal of the American Oil Chemists' Society*, **48** 794.

[8] Wickbold, R. (1974) Analytische Beiträge zum biologischen Abbau von Tensiden. *Tenside Detergents*, **11** 137.

[9] Brüschweiler, H., Gämperle, H. and Schwager, F. (1983) Primary biodegradation, ultimate biodegradation, and biodegradation intermediates of alkylphenol ethoxylates. *Tenside*, **20** 317.

[10] Saeger, V.W., Kuehnel, R.G. Linck, C. and Gledhill W.E. (1980) *Biodegradation of Sterox and Dimersol olefin-derived alkylphenol ethoxylates*, Report No. ES-80-SS-46, Monsanto Industrial Chemicals Company, St. Louis, MO.

[11] Brüschweiler, H. and Gämperle, H. (1982) Primary and ultimate biodegradation of alkylphenol ethoxylates, in *XIII Jornadas Com.* vol. 13, Español Deterg., Barcelona, p. 55.

[12] Yoshimura, K. (1986) Biodegradation and fish toxicity of nonionic surfactants. *Journal of the American Oil Chemists' Society*, **63** 1590.

[13] Hughes, A.I, Peterson, D.R. and Markarian, R.K. (1989) Comparative biodegradability of linear and branched alcohol ethoxylates, in *Proceedings of the American Oil Chemists' Society Annual Meeting*, AOCS Press, Champaign.

[14] Kravetz, L., Salanitro, J.P., Dorn, P.B., Guin, K.F. (1991) Influence of hydrophobe type and extent of branching on environmental response factors of nonionic surfactants. *Journal of the American Oil Chemists' Society*, **68** 610.

[15] Varineau, P.T. and Williams, J.B. (1997) The aerobic biodegradation of nonylphenol, nonylphenol ethoxylates and their biodegradation intermediates, in *Proceedings of the AOCS World Conference on Environmental Challenges, Brussels, March 1997* (in prep.).

[16] Brown, D. (1988) Alkylphenol ethoxylates: an environmental impact assessment, in *Proceedings of the 2nd World Surfactants Congress of CESIO in Paris, May 1988*, CESIO, Brussels.

[17] Stiff, M.J., Rootham, R.C. and Culley, G.E. (1973) The effect of temperature on the removal of nonionic surfactants during small-scale activated-sludge sewage treatment. *Water Research*, **7** 1003.

[18] Birch, R.R. (1991) Prediction of the fate of detergent chemicals during sewage treatment. *Journal of Chemical Technology and Biotechnology*, **50** 411.

[19] Birch, R.R. (1991) Recent developments in the biodegradability testing of nonionic surfactants. *Riv. Ital. Sost. Grasse*, **68** 433.

[20] Talmage, S. (1994) *Environmental and Human Safety of Major Surfactants*, The Soap and Detergent Association, Lewis Publishers, Boca Raton.

[21] Ekelund, R., Granmo, A. Magnusson, K. Berggren, M. and Bergman, A. (1993) Biodegradation of 4-nonylphenol in seawater and sediment. *Environmental Pollution*, **79** 59.

[22] Lee, H.B. and Peart, T.E. (1995) Determination of 4-nonylphenol in effluent and sludge from sewage treatment plants. *Analytical Chemistry*, **67** (13) 1976.

[23] Ball, H.A. Reinhard, M. and McCarty, P.L. (1989) Biotransformation of halogenated and non-halogenated octylphenol polyethoxylate residues under aerobic and anaerobic conditions. *Environmental Science and Technology*, **23** 951.

[24] Williams, J.B., Blessing, R.L. and Varineau, P.T. (1996) *Aquatic fate and effects testing: data on alkylphenol ether carboxylates*, Project Report of Union Carbide Corporation, South Charleston.

[25] Field, J.A. and Reed, R.L. (1996) Nonylphenol polyethoxy carboxylate metabolites of nonionic surfactants in US paper mill effluents, municipal sewage treatment plant effluents, and river waters. *Environmental Science and Technology*, **30** (12) 3544-50.

[26] Naylor, C.G., Williams, J.B., Varineau, P.T. and Webb, D.A. (1996) Nonylphenol ethoxylates in an industrial river, in *Proceedings of the 4th World Surfactants Congress of CESIO*, CESIO, Brussels, pp. 4, 378.

[27] Giger, W., Ahel, M. and Koch, M. (1986) Behaviour of alkylphenol polyethoxylate surfactants in mechanical–biological sewage treatment. *Vom Wasser*, **67** 69.

[28] Naylor, C.G., Mieure, J.P., Adams, W.J., Weeks, J.A., Castaldi, F.J., Ogle, L.D. and Romano, R.R. (1992) Alkylphenol ethoxylates in the environment. *Journal of the American Oil Chemists' Society*, **69** (7) 695.

[29] Ahel, M., Giger, W. and Koch, M. (1994) Behaviour of alkylphenol polyethoxylate surfactants in the aquatic environment I. Occurrence and transformation in sewage treatment. *Water Research*, **28** (5) 1131.

[30] Weeks, J.A., Adams, W.J. Guiney, P.D. Hall, J.F. and Naylor, C.G. (1996) Risk assessment of nonylphenol and its ethoxylates in US river water and sediment, in *Proceedings of the 4th World Surfactants Congress of CESIO*, CESIO, Brussels, p. 3, 276.

[31] Naylor, C.G. (1995) Environmental fate and safety on nonylphenol ethoxylates. *Textile Chemist and Colorist*, **27** (4) 29.

[32] Blackburn, M.A. and Waldcock, M.J. (1995) Concentrations of alkyl phenols in rivers and estuaries in England and Wales. *Water Research*, **29** (7) 1623.

[33] Clark, L.B., Rosen, R.T., Hartman, T.G., Louis, J.B., Suffer, I.H., Lippincott, R.L. and Rosen, J.D. (1991) Determination of nonregulated pollutants in three New Jersey publicly owned treatment works (POTWs). *Research Journal WPCF*, **63** (2) 104.

[34] Crescenzi, C., Coorcia, A.D., Samperi, R. and Marcomini, A. (1995) Determination of nonionic polyethoxylate surfactants in environmental waters by liquid chromatography/ electronspray mass spectrometry, *Analytical Chemistry*, **67** (11) 1797.

[35] Ahel, M. and Giger, W. (1993) Partitioning of alkylphenols and alkylphenol polyethoxylates between water and organic solvents. *Chemosphere*, **26** 1471.

[36] Ahel, M., Giger, W. and Schaffner C. (1996) Behaviour of alkylphenol polyethoxylate surfactants in the aquatic environment III. Occurrence and elimination of their persistent metabolites during infiltration of river water to groundwater. *Water Research*, **30** (1) 37.

[37] Giger, W., Brunner, P.H. and Schaffner, C. (1984) 4-Nonylphenol in sewage sludge: accumulation of toxic metabolites from nonionic surfactants. *Science*, **225** 623.

[38] Brunner, P.H., Capri, S., Marcomini, A. and Giger, W. (1988) Occurrence and behaviour of linear alkylbenzenesulphonate, nonylphenol, nonylphenol mono- and diethoxylates in sewage and sludge treatment. *Water Research*, **22** 1465.

[39] Marcomini, A., Capri, S., Brunner, P.H. and Giger, W. (1988) Mass fluxes of linear alkylbenzenesulphonates, nonylphenol, nonylphenol mono- and diethoxylate through a sewage treatment plant. *Commun. Eur. Comm. Rep.*, **11350** 266.

[40] Ahel, M. Giger, W. and Schaffner C. (1994) Behaviour of alkylphenol polyethoxylate surfactants in the aquatic environment II. Occurrence and transformation in rivers. *Water Research*, **28** (5) 1143.

[41] Ejlertsson, J. Öquist, M. and Svensson, B.H. (1996) *Anaerobic degradation of mono and nonylphenol diethoxylates in sludge and landfilled sludge and municipal solid waste*, Department of Water and Environmental Studies, University Linköping, Sweden.

[42] Marcomini, A., Capel, P.D., Lichtensteiger, T., Brunner, P.H. and Giger, W. (1989) Behaviour of aromatic surfactants and PCBs in sludge-treated soil and landfills. *Journal of Environmental Quality*, **18** 523.

[43] Kalbfus, W. and Zellner, A. (1995) *Belastung bayerischer Gewässer durch Nonylphenole. Stoffe mit endokriner Wirkung im Wasser*, Oldenbourg Verlag, Munich, pp. 55-65.

[44] Giger, W. (1995) Spurenstoffe in der Umwelt. *EAWAG News 40D*, November.

[45] Trocmé, M., Tarradellas, J. and Védy, J.-C. (1988) Biotoxicity and persistence of nonylphenol during incubation in a compost–sandstone mixture. *Biology and Fertility of Soils*, **5** 299-303.

[46] Hughes, A.I., Fisher, J.J. and Brumbaugh, E.H. (1996) Biodegradation of NPE in soil, in *Proceedings of the 4th World Surfactants Congress of CESIO*, CESIO, Brussels, pp. 4, 365.

[47] Ahel, M. (1991) Infiltration of organic pollutants into groundwater: field studies in the alluvial aquifer of the Sava River. *Bulletin of Environmental Contamination and Toxicity*, **47** 586.

[48] Schaffner, C., Ahel, M. and Giger, W. (1987) Field studies on the behaviour of organic micropollutants during infiltration of river water to ground water. *Water Science and Technology*, **19** 1195.

[49] Dorn, P.B., Salanitro, J.P., Evans, S.H. and Kravetz, L. (1993) Assessing the aquatic hazard of some linear and branched nonionic surfactants by biodegradation and toxicity. *Environmental Toxicology and Chemistry*, **12** 1751.

[50] Comber, M.H.I., Williams, T.D. and Stewart, K.M. (1993) The effects of nonylphenol on *Daphnia magna*. *Water Research*, **27** 273.

[51] CMA (1990) *Acute flow-thru toxicity of nonylphenol to the marine alga* (Skelatonema costatum), Contractor's Report, Chemical Manufacturers' Association, Washington, DC.

[52] CMA (1991) *Early life-stage toxicity of nonylphenol to the fathead minnow* (Pimephales promelas) Contractor's Report, Chemical Manufacturers' Association, Washington, DC.

[53] CMA (1991) *Chronic toxicity of nonylphenol to the mysid shrimp* (Mysidopsis bahia), Contractor's Report, Chemical Manufacturers' Association, Washington, DC.

[54] CMA (1994) *Chronic toxicity of nonylphenol to the freshwater invertebrate* (Ceriodaphnia dubia), Contractor's Report, Chemical Manufacturers' Association, Washington, DC.

[55] CMA (1990) *Acute static toxicity of nonylphenol to the freshwater alga* (Selenastrum capricornutum), Contractor's Report, Chemical Manufacturers' Association, Washington, DC.

[56] Gobas, F.A.P.C. (1993) A model for predicting the bioaccumulation of hydrophobic organic chemicals in aquatic food-webs: application to Lake Ontario. *Ecological Modeling*, **69** 1.

[57] Ekelund, R., Bergman, A., Granmo, A. and Berggren, M. (1990) Bioaccumulation of 4-nonylphenol in marine animals—a re-evaluation. *Environmental Pollution*, **64** 107.

[58] McLeese, D.W., Sergeant, D.B., Metcalfe, C.D., Zitko, V. and Burridge, L.E. (1980) Uptake and excretion of aminocarb, nonylphenol and pesticide diluent 585 by mussels (*Mytilus edulis*). *Bulletin of Environmental Contamination and Toxicology*, **24** 575.

[59] McLeese, D.W., Sergeant, D.B., Metcalfe, C.D., Zitko, V. and Burridge, L.E. (1981) Lethality and accumulation of alkylphenols in aquatic fauna. *Chemosphere*, **10** 723.

[60] CMA (1996) *Research Summary*, APE Panel, Chemical Manufacturers' Association, Washington, DC.

[61] Ahel, M., McEvoy, J. and Giger, W. (1993) Bioaccumulation of the lipophilic metabolites of nonionic surfactants in freshwater organisms. *Environmental Pollution*, **79** 243.

[62] Nimrod, A.C. and Benson, W.H. (1996) The estrogenic effects of alkylphenol ethoxylates: a literature review submitted to CMA. *Critical Reviews in Toxicology*, **26** (3) 335.

[63] Mueller, G.C. and Kim, U. (1978) Displacement of estradiol from estrogen receptors by simple alkyl phenols. *Endocrinology*, **102** 1429.

[64] Jobling, S. and Sumpter, J.P. (1993) Detergent components in sewage effluents are weakly oestrogenic to fish. An *in vitro* study using rainbow trout (*Oncorhynchus mykiss*) hepatocytes. *Aquatic Toxicology*, **27** 361.

[65] White, R., Sumpter, S., Hoare, S.A. and Parker, M.G. (1994) Environmentally persistent alkylphenolic compounds are estrogenic. *Endocrinology*, **135** 175.

[66] Jobling, S., Sheahan, D., Osborne, J.A., Matthiessen, P. and Sumpter J.P. (1996) Inhibition of testicular growth in rainbow trout (*Oncorhynchus mykiss*) exposed to estrogenic alkylphenolic chemicals. *Environmental Toxicology and Chemistry*, **15** 194.

2 Surfactants in environmental remediation: aquifer restoration

J.H. Harwell, R.C. Knox and D.A. Sabatini

2.1 Introduction

2.1.1 Groundwater contamination

Groundwater contamination by hydrocarbon, and especially chlorohydrocarbon liquids, is one of the most intractable and expensive types of pollution to remediate. The availability of a large selection of non-toxic surfactants widely used in foods, food packaging or food preparation areas, together with their known biodegradability, and the results of two decades of research into the use of surfactants in tertiary oil recovery, is producing a great deal of excitement in the environmental community. Within the last 2 years there have been several successful field tests to demonstrate the technical feasibility of using surfactants to remediate solvent- and fuel-contaminated aquifers. This chapter discusses the size and background of the environmental problem of groundwater contamination, the manner in which surfactants are being used to address the problem and areas of research still remaining to be addressed.

As the end of the twentieth century approaches, government, industry, and the general public are very conscious of the sensitivity of the environment to pollution arising from human activities. There are broad efforts world wide both to prevent future pollution and to remediate past mistakes and excesses.

Nowhere are the problems of the past more recalcitrant than in the case of aquifer contamination by non-aqueous phase liquids (NAPLs). One half of all drinking water in the US comes from ground water. NAPL contamination of aquifers is particularly pernicious because liquids immiscible with water tend to dissolve into water only at very low levels; if, however, these levels exceed those deemed safe for human consumption, then small volumes of an NAPL in an aquifer can contaminate millions of cubic metres of ground water [1, 2]. While surfactants may play an important role in such environmental technologies as *ex situ* soil washing [3], adsorbing barriers [4] and wastewater treatment [5], the area in which they are most likely to have a major impact is in groundwater remediation where subsurface NAPLs constitute the contaminant.

There are two important types of NAPL contamination. Certainly the most common is LNAPL or light non-aqueous phase liquid contamination. LNAPL contamination is common near gasoline, jet fuel and diesel fuel manufacturing, storage and refueling sites, for example. It sometimes also occurs around oil wells and oil storage, processing and transportation facilities. A frequently seen scenario involves the leakage of fuel from an underground storage tank (UST). This occurs years after a tank has been installed in a subsurface location, once the tank has corroded to the point where it begins to leak. The leak will probably go unnoticed until after thousands of litres of fuel have been lost. If the fuel leakage is of sufficient volume that it cannot be contained by the 'sorptive' capacity of the soil, then the LNAPL may reach the water table, where it will form a pool or lens, floating on the aquifer (Figure 2.1). As the water table rises and falls seasonally, fuel droplets become distributed both above and below the water table, trapped in the porous medium formed by the sand grains by the high surface tension between the NAPL and the aquifer water (Figure 2.2). High interfacial tension keeps the oil drop inside the pores of the sand and thus prevents the oil from being removed by pumping or buoyant forces; thus, it remains in contact with the ground water, where low levels of dissolution continually contaminate the ground water [6]. There is a daunting number of such sites; an estimated 50 000 UST sites in the US alone. To this number must be added additional sites where there is groundwater contamination because of spills, pipeline leaks, etc., taking the total number of sites into the hundreds of thousands [7]. At each site it is not only the water immediately around the spill itself that becomes contaminated. Because ground water flows under the ground in the same manner as surface water flows in rivers, a single spill can, over the years, result in the contamination of ground water many kilometres away as the dissolved contaminant migrates with the flowing ground water (Figure 2.3). If the ground water

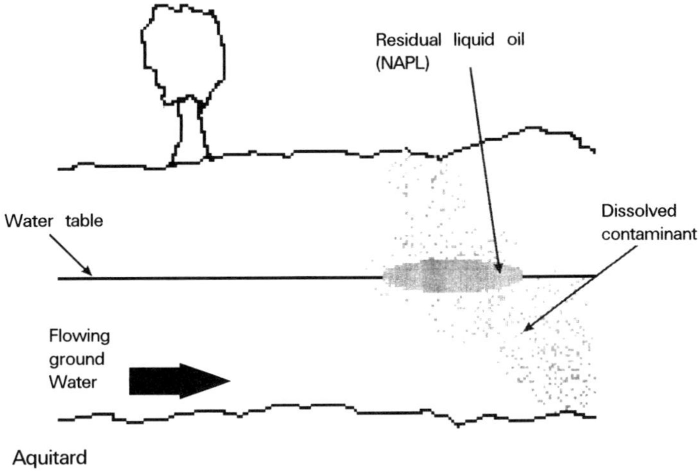

Figure 2.1 Lens of NAPL floating on ground water at the water table.

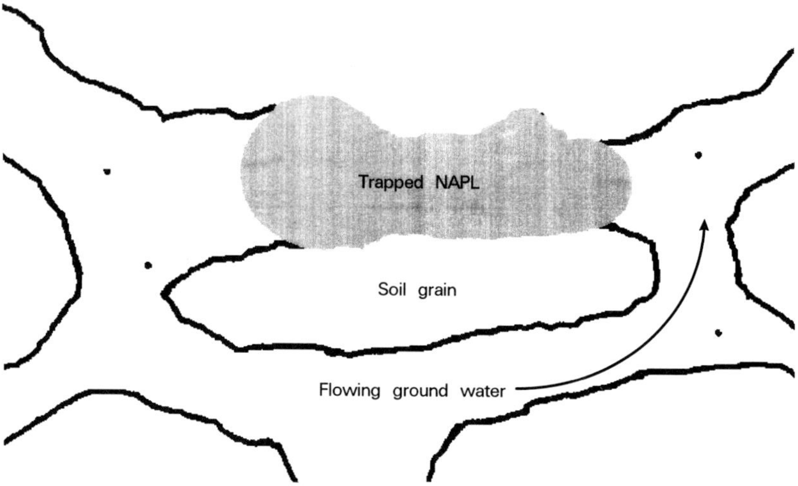

Figure 2.2 Droplet of NAPL trapped by capillary forces between aquifer sand grains.

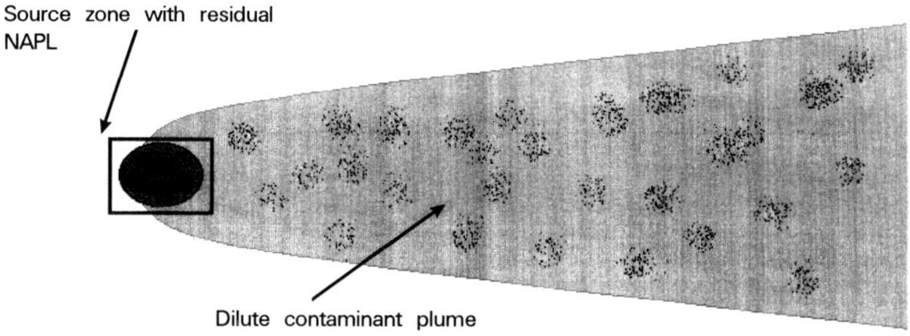

Figure 2.3 Plume of dissolved contaminant spreading from NAPL spill site.

communicates with a river or lake, the surface water of the river or lake may become contaminated by a subsurface source many kilometres away

As pernicious as LNAPL contamination may be, many LNAPL sites, especially those where the water table is not too deep, can be remediated by excavation, air sparging or soil vapor extraction. The case of dense non-aqueous phase liquid (DNAPL) contamination is much worse. When solvents like trichloroethylene, orthodichlorobenzene or perchloroethylene are released into the environment in large volumes, the solvent does not float on top of the aquifer as does the LNAPL. DNAPLs have a density greater than water; when the spilled liquid reaches the water table, it continues to migrate downwards and can penetrate to the aquitard (the impermeable layer that prevents the ground water from migrating more deeply into the ground) (Figure 2.4). As with the LNAPL,

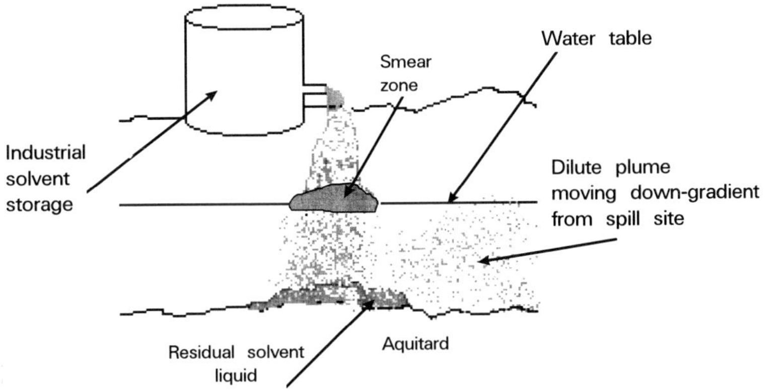

Figure 2.4 DNAPL penetrates the aquifer below the water table.

however, droplets of the migrating DNAPL become trapped by capillary forces in the porous medium of the aquifer's solid matrix, forming discrete droplets because of the interfacial tension between the DNAPL and the ground water (Figure 2.2). At over half of the superfund sites in the US (those sites identified as the most highly contaminated and having the greatest priority for clean-up) the principal problem is DNAPL contamination of the ground water. As with LNAPL, the flow of ground water from the site of the spill results in millions of litres of down-gradient water becoming contaminated.

DNAPL contamination has been identified as a priority for advanced remediation technology development by the US Environmental Protection Agency (EPA), Department of Defense (DOD) and Department of Energy (DOE). The EPA's 1995 Innovative Treatment Technologies Database [8] lists 205 test sites with halogenated volatile contaminants, 47 with halogenated semivolatiles and 26 with polychlorinated biphenyls (PCBs). A very large number of DOD and DOE sites have been contaminated by DNAPLS. The desire of the US Government to reduce the number of operating military bases both within the US and around the world has been hampered to a large degree by the liability associated with turning over to civilian authorities sites that have high levels of groundwater contamination. In some locations the US military has a policy of compensating local governments for the cost of remediating a contaminated site that is to be turned over to the host government. The world-wide liability for these sites runs into billions of US dollars and is a major incentive for developing more cost-effective remediation technologies. Beyond these government sites, there are large numbers of industrial sites where DNAPL contamination exists; such sites include not only chemical plants but also electronics, dry cleaning and metal fabrication facilities, to list only a few types. There is presently no accepted technology for remediating ground water contaminated by DNAPL. Clean-up of these sites by any means currently envisioned will cost many billions of US dollars. It has been estimated that the cost of remediation of US DOD and DOE sites alone may reach $325 billion (1995 dollars) by 2070; the US Departments of Interior and Agriculture report an additional 29000 sites, including abandoned oil and gas fields, which may cost an additional $10 billion to remediate [7].

While surfactants have a significant potential for aiding in the remediation of LNAPL contaminated sites, it is in the area of DNAPL contamination where there is the most enthusiasm for surfactants in environmental remediation [9]. Surfactants, through their ability to solubilize non-aqueous liquids into water and to reduce interfacial tensions, have great promise for aiding in the remediation of such sites. Additionally, there are large numbers of biodegradable, non-toxic surfactants that are commercially produced in sufficient quantities and at low enough cost for utilization in groundwater remediation. Finally, modern surfactant technology has a great depth of understanding regarding both surfactant science and the technology of cleaning, which can be applied to this problem.

2.1.2 Surfactant studies

The literature on surfactant flushing of contaminants is already extensive and is growing at a rapid pace [10]. The basic mechanisms by which surfactants may be used to flush contaminants from aquifers are discussed by West and Harwell [11] and Sabatini *et al.* [12]. Vignon and Rubin [13] discuss selection of surfactants for solubilizing anthracene and biphenyl with alkyl and alkylphenol polyethoxylates. Abdul *et al.* [14] discuss the use of different surfactant types for the solubilization of automatic transmission fluid contaminating a sandy aquifer. Valsaraj and Thibodeaux [15] have evaluated partition coefficients for eleven organic compounds in sodium dodecyl sulfate micelles.

Edwards *et al.* [16] have examined the use of polyethoxylates for solubilizing polynuclear aromatic hydrocarbons (PAHs). West [17] has also utilized nonionic surfactants in studying the solubilization of perchloroethylene (PCE), trichloroethylene (TCE) and 1,2-dichloroethylene (1,2-DCE). Rouse *et al.* [18, 19] have compared monosulfonates and disulfonates for solubilization while also examining the tendency of these different surfactant types to precipitate or adsorb; they conclude that the disulfonates have important advantages in reducing surfactant loses from precipitation and adsorption, while giving very similar solubilizations. Shiau *et al.* [20] have examined the use of surfactants with direct food additive status for solubilization and for microemulsion formulation, demonstrating both the feasibility of using food additives in aquifer remediation and the greatly enhanced contaminant solubilizations observed with Winsor Type III microemulsions. Pennel *et al.* [21] have examined the behavior of surfactants in removing dodecane from one-dimensional soil columns, with special emphasis on the role of interphase mass transfer resistance in the performance of surfactant flushes using micellar solubilization as the contaminant removal mechanism. Krebs-Yuill *et al.* [22] conclude that surfactant recovery and reuse is critical to the economics of surfactant-enhanced aquifer remediation.

2.2 Overview of the potential of surfactant-enhanced aquifer remediation

It is neither practical nor advantageous to consider using surfactants for the dilute plume of contaminants that extends from the source of residual contamination in an NAPL contaminated aquifer [22]; rather, the goal of surfactant-enhanced aquifer remediation is to rapidly and economically remove the residual liquid contaminant that is the source of the dissolved plume, which can extend several kilometres down-gradient from the site of the original NAPL release (Figure 2.3). The approach of focusing the remediation on the zone where the origin of the contamination lies is called source-zone treatment. Surfactants can be used for enhanced remediation in aquifers using two separate remediation mechanisms, both focused on the source zone. The ideal site for application of surfactant-enhanced aquifer remediation is one where removal of the source zone allows the plume of dissolved contaminant to be ignored, once its growth is cut off.

It is possible to reduce dramatically the time and cost of a source-zone remediation project using surfactants simply to *solubilize* the residual liquid contaminants into normal aqueous micelles (Figure 2.5). When the solubilization mechanism is used, all that is necessary for enhancement to occur is that micelles be propagated through the zone of residual contamination (Figure 2.6). A more time-efficient approach is to *mobilize* the trapped NAPL droplets out of the pores. This requires selection of the surfactant system to produce an ultra-low

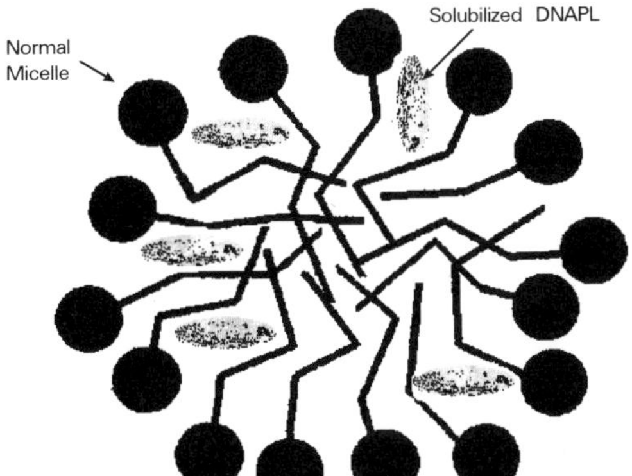

Figure 2.5 NAPL molecules solubilized in normal micelle.

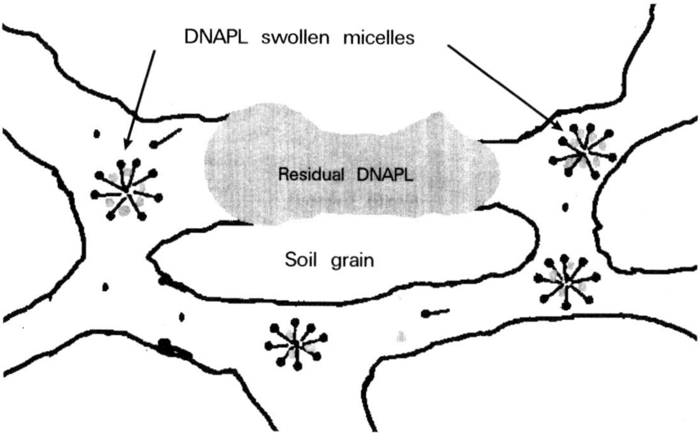

Figure 2.6 NAPL-swollen micelles increase contaminant removal rate as water is pumped to the surface.

Table 2.1 Comparison of solubilization and mobilization as remediation technologies

Flushing technology	Contaminant removal mechanism	Advantages	Disadvantages
Solubilization	Solubilization in normal micelles	Very robust Relatively independent of surfactant type Little concern with vertical migration of residual liquid contaminant	Low enhancement rates relative to mobilization Remediation time excessive when residual saturation exceeds a few volume percent May cause fractionation of mixed NAPL, leaving insoluble materials behind
Mobilization	Residual liquid released from soil pores by low interfacial tensions seen in Winsor Type III/IV microemulsions	Remediation time independent of residual saturation Much higher solubilizations than seen with micelles alone Uses no more surfactant than solubilization mechanism Surfactant recovery and reuse may not be necessary in well designed, low pore volume injection with mobility control (EOR type process)	Vertical migration of residual liquids may complicate aquifer remediation More sensitive to changes in intensive thermodynamic variables than solubilization Not all surfactants can make Type III microemulsions with many contaminant liquids May require use of low molecular weight alcohol to prevent liquid crystal formation

interfacial tension between the injected surfactant solution and the residual NAPL. The ultra-low interfacial tension reduces the capillary force that traps the residual droplets in the aquifer, thus releasing the trapped drops from the pores between the sand grains and allowing the NAPL to be pumped directly to the surface. This process is referred to as mobilization; it requires a high adsorption of surfactant monomers at the NAPL/water interface (Figure 2.7). Each mechanism has various advantages and disadvantages (Table 2.1), which must be evaluated for each contaminated site before deciding which mechanism to use.

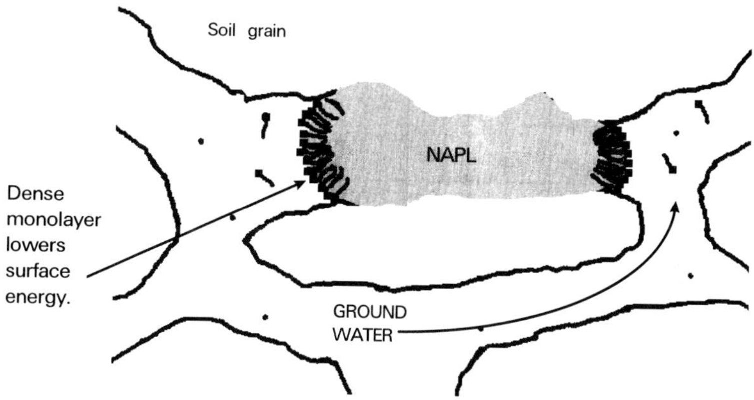

Figure 2.7 High Gibbs adsorption at NAPL/ground water interface lowers interfacial tension.

As one might anticipate, there are a great many similarities between the problem of NAPL remediation and the problem of enhanced oil recovery (EOR). In EOR, large residual saturations of petroleum crude are trapped in oil reservoirs by surface tension forces after water flooding has been used to extend the productive life of a producing oil field [23]. Following the energy crisis of the 1970s there was an enormous effort in many nations to develop surfactant systems for recovering a substantial fraction of this trapped oil. While the technology that was developed was successful at recovering the oil, the general drop in oil prices in the 1980s, together with the continued uncertainty of future oil prices in the 1990s, has prevented widespread application of this technology. EOR technology focused exclusively on low interfacial tension systems that would mobilize the residual oil; the goal was to produce an oil bank that would move ahead of the injected surfactant, effectively sweeping out remaining oil. No EOR research ever examined the feasibility of using solubilization because of the high residual saturations involved and the associated problem of separating the oil from the micelles.

A great many lessons learned in the development of surfactant EOR technology are directly applicable to groundwater remediation [24]. A principal concern in the design of a surfactant-based aquifer remediation project must be the phase behavior of the surfactant at the conditions in the aquifer [18]. Electrolyte concentrations and temperature must be considered in choosing the surfactant to be used. Many important commercial surfactants will not remain active at aquifer conditions of temperature and electrolyte concentrations; rather, they can be expected to precipitate or form a coacervate phase. Coacervation must also take into account the effect of the presence of the NAPL on the tendency of the surfactant to form a coacervate phase. Surfactants that form an isotropic phase at aquifer conditions may still form liquid crystals in the presence of the NAPL.

As with surfactant EOR, the cost of the surfactant is a major portion of a surfactant-based groundwater remediation project [22, 25]. For this reason, it is important to the overall process economics both that the surfactant be modestly priced and that the adsorption of the surfactant on the aquifer matrix be minimized [18]. Losses of surfactant to adsorption can essentially double the surfactant costs in a groundwater remediation project and the surfactant costs can be the single greatest cost element in the project. Another major problem in both technologies is concern about the heterogeneous flow field in the treatment zone. There is a tendency for injected fluids to follow paths of high permeability, leaving low permeability zones largely untreated [26–29]. As the injected surfactant removes the oil or NAPL from the treated zones, the permeability of these zones to water actually increases, resulting in still greater flow heterogeneity. This problem was addressed in EOR by the use of polymers to increase the density and viscosity of the injected fluids. Such an approach may also prove worthwhile in NAPL remediation [24], although this technology had not been field tested for aquifer remediation at the time of writing.

For all the similarities between aquifer remediation and EOR, there are significant differences. Perhaps the most obvious is that surfactant toxicity and biodegradability become primary concerns in aquifer remediation, whereas they can be neglected in an oil reservoir; fortunately, most commercial surfactant types have already been established to be non-toxic to mammals and biodegradable, at least in municipal sewage treatment facilities. Beyond this, however, a number of surfactant types, such as the ethoxylated sorbitol fatty esters, are actually edible, being approved in various countries as direct food additives [30–32]. An even larger number of surfactants are classified as indirect food additives, usable in food preparation areas and in food packaging [33]. These facts may prove crucially important in obtaining regulatory approval for the injection of surfactants into groundwater aquifers; understandably, there is great concern and considerable amounts of prohibitive legislation regarding the injection of any chemicals into ground water. To date, however, there has been a great deal of willingness in the US by government officials responsible for enforcing ground water protection legislation to allow injection of commercial surfactants into already contaminated aquifers in the interests of developing new remediation technology; in all such cases the injected surfactant has been known to be biodegradable by standard tests and to possess at least indirect food additive status under US Food and Drug Administration regulations [34–36]. It is unfortunate that there is still relatively little known about surfactant biodegradation either under general aquifer conditions or under the anaerobic conditions that might be encountered during an aquifer remediation project. As the technical and economical feasibility of surfactant-enhanced aquifer remediation is demonstrated, the topic of surfactant biodegradation in ground water will be of growing importance [37].

Perhaps the most critical difference between surfactant-enhanced aquifer remediation and EOR is the difference in process effectiveness required before a project can be regarded as a success. In EOR, a project that recovers more than 10% of residual oil-in-place might well be an economic success. A project that recovers 90% of the residual oil-in-place would be an astounding technical success. For surfactant-enhanced aquifer remediation, the criteria of success are not generally accepted and may have to be defined on a site-by-site basis. Ultimately, however, an aquifer restoration project can only be considered successful if the level of restoration is such that the risk to human health and the environment is sufficient for the site to be considered 'restored'. How much contaminant removal this implies may vary widely, depending on the nature of the contaminant, the location of the contamination and the location and utilization of the contaminated aquifer. An aquifer restoration project that recovered 99.9% of the trichloroethylene from a contaminated aquifer used for municipal drinking water, but which still left contaminant concentrations in the ground water above drinking water levels, which may defined in the ppb range, might be considered to be a costly failure. In contrast, a restoration project that removed only 50% of the trichloroethylene from a brackish aquifer that fed into a river might be considered a success if the flux of contaminant away from the site and into the river was reduced below hazardous levels and it was concluded that subsequent contaminant migrating away from the site would either biodegrade, adsorb or become sufficiently dilute that there was no realistic possibility that it might be considered hazardous by the time it reached any conceivable receiver. The authors believe this means that future potential surfactant-enhanced aquifer remediation projects will need to be evaluated in the light of the risk-based corrective action (RBCA) policy. The principle behind RBCA (pronounced 'Rebecca') is that even when the environment cannot be restored to pristine conditions there may be a level of restoration that is both economically feasible (without bankrupting all parties involved, including governmental parties) and will result in the substantial elimination of future dangers associated with the site. An EOR project could be evaluated based on whether or not it would be a financial success; a surfactant-enhanced aquifer remediation project will have to be evaluated at least partly on the willingness of governmental regulators, and therefore the public in general, to consider an aquifer 'restored' even if it has not been returned to pristine condition.

There are also some differences between surfactant-enhanced aquifer remediation and surfactant EOR that actually make aquifer remediation much easier, in a technical sense, than EOR operations. For example, aquifer temperatures and electrolyte concentrations are not as extreme as those found in oil reservoirs, which makes a wider variety of chemical types of surfactants acceptable for consideration in aquifer remediation projects. The most important difference, however, is certainly the cost of wells and the consequent reduction in well spacings [22]. In EOR one seldom encounters wells less than 1 km in depth, and wells at several times that depth are not at all unusual. In aquifer remediation, a depth of 50 m constitutes a deep aquifer; many important applications of surfactant-enhanced aquifer remediation occur at depths of less that 10 m. These differences in orders of magnitude in depth translate into differences in orders of magnitude in well spacings; as a result, the design of an aquifer remediation project can afford to use as many wells as necessary to reduce the time of the project and to minimize the effects of subsurface heterogeneities.

Somewhat unexpectedly, the simple difference of being able to have closer well spacings opens up a wide variety of alternatives for the design of microemulsions for use in aquifer remediation by mobilization. In EOR, well spacings are frequently so large that propagation of an injected chemical from the injection wells to the producing wells may take years. During this time, the money spent to purchase the surfactant used to increase the oil recovery constitutes an investment that is not producing any revenue. A direct consequence of this is that the economics of EOR require the amount of surfactant injected into the reservoir to be minimized. The goal of the most sophisticated EOR technology was to use a volume of surfactant solution that represented as small a fraction of the liquid volume of the reservoir (a pore volume) as possible [23, 24]. The use of small pore volume 'slugs' of injected surfactant not only exacerbated the effects of flow heterogeneities in the reservoir but also resulted in chromatographic separation of dissimilar surface active species used to formulate the injected mixture. If the injected mixture had to be mono-isomeric or had to be formulated without co-surfactants such as low molecular weight alcohols, then this further required that the surfactant molecule that was chosen be designed specifically to produce an ultra low interfacial tension, such as is only seen in Winsor Type III and Type IV microemulsions [38], between the reservoir water and the petroleum crude to be recovered. These restrictions further increased the cost of the surfactants.

With surfactant-enhanced aquifer remediation, however, the reduced cost of wells allows a closer well spacing and consequently reduces the time required to propagate an injected solution from injection to extraction wells. The impact of the cost of the surfactant is sufficiently reduced that it is no longer economically inconceivable to saturate the contaminated zone of the aquifer with surfactant.

Further, once this possibility is realized, then the problem of flow heterogeneity is reduced and the problem of chromatographic separation is practically (if not quite completely) eliminated. Once mixtures of surfactants can be used to design the active mixture, a wider variety of approaches becomes possible for obtaining a Winsor Type III or middle-phase microemulsion [39] between the injected surfactant solution and the NAPL contaminating the aquifer. The use of low molecular weight alcohols and hydrotropes together with the surfactants in the active mixture can also be considered without worrying about chromatographic separation.

A further consequence of the use of large pore volume chemical slugs is that recovery and reuse of the surfactant in the remediation operation becomes a viable way of reducing the total cost of the surfactants in the project [22, 25]. Rather than buying sufficient surfactant to flush the contaminated zone four or five times, only enough to flush the zone a little over once need be purchased; this surfactant is then regenerated when recovered and reinjected. While this procedure is technically feasible [40, 41], it will require even greater flexibility on the part of government regulators if it is to be used in actually cleaning up some of the world's most dangerously polluted sites. If both the public and those responsible for overseeing the protection of the environment can be assured that reinjection of the decontaminated surfactant is both safe and desirable, in order to allow more effective use of the limited resources available for environmental remediation, then significant progress can be made in the immediate future towards reducing the danger to the environment of past environmental sins.

2.3 Recent results

Though surfactant-enhanced aquifer remediation has been the subject of much recent research [10], at the time of writing only a handful of reports of field tests of the technology have appeared in the refereed literature. Abdul *et al.* [14] reported a field test in a 10' by 5' pilot for removing PCB contamination from the unsaturated zone. Non-ionic surfactants were used. After six pore volumes, an estimated 10% of the contaminant had been removed from the site. Two other tests with non-ionic surfactants, in Borden, Ontario [42] and Corpus Christi, Texas [43], reported the use of surfactant to remove two DNAPLs, PCE and carbon tetrachloride, respectively. In both cases good solubilization of the contaminant was reported. The US EPA's 1995 Innovative Treatment Technologies Database [8], however, lists only two surfactant flushing projects, one in the predesign stage and the other in the design stage.

More recently,within the last two years, there have been reports of very successful demonstrations of surfactant-enhanced aquifer remediation; although these results have not yet appeared in the refereed literature, they have been presented at a national meeting of the American Chemical Society (San Francisco,

April 1997). In one of these reports, Brown *et al.* [44], the anionic surfactant sodium dihexyl sulfosuccinate was injected into the lower region of an alluvial aquifer at Hill Air Force Base, near Ogden, Utah. The aquifer was contained on its sides and below by a thick clay layer. The deepest part of the aquifer was approximately 17 m below the surface of the ground. The treated zone had an initial average residual DNAPL saturation of approximately 4%, while the lower 1.6 m of the treated zone had a residual saturation of DNAPL approaching 20%. The DNAPL consisted of approximately 95% trichloroethene, trichloroethane and tetrachloroethene. The solubility of the DNAPL in the aquifer water was estimated at 1100 mg/l. Addition of 8% surfactant raised the solubility to 60 000 mg/l (an increase of a factor of 55); addition of 7000mg/l of NaCl and 4% isopropanol to the surfactant raised the solubility of the DNAPL to over 600 000 mg/l (an increase of a factor of over 500). By the use of partitioning tracer tests it was determined that after injection of 2.4 pore volumes of the active mixture, over a period of 3 weeks, more than 99% of the DNAPL was removed, leaving a residual saturation of only 0.04%. Even though the monitoring well system in the treated zone was not isolated from the rest of the contaminated aquifer, the DNAPL concentration in the central monitoring well dropped to 30 mg/l. Overall, the injection of the 2.4 pore volumes of surfactant, without use of mobility control agents, removed 510 of the original 520 gallons of DNAPL in the treated zone.

Another test of the same surfactant at the same site was completed in April, 1997, but air injection was used to generate an *in situ* foam; consequently, the isopropanol was omitted from the active mixture, as it is a foam negative factor. A preliminary analysis of the results was presented at a Remediation Technology Development Forum (RTDF) held at Hill AFB, Ogden, Utah, in May 1997. The purpose of the foam was to improve the sweep efficiency of the surfactant mixture by blocking preferentially the most permeable zones of the contaminated region. While the use of mobility control was expected to improve sweep efficiency, removal of the alcohol also reduced the solubilization capacity of the active mixture. The results have not yet been fully analyzed, much less reported, but they appear to be comparable with the earlier test, while having used less surfactant. These early, incomplete reports indicated that a stable, *in situ* foam was successfully generated and propagated from the injection to the extraction wells. This foam injection project, on which the chief scientists were Dr Clarence Miller and Dr George Hirasaki of Rice University, Houston, Texas, will be reported as part of the results for the Advanced Applied Technology Development Facility (AATDF) Project funded at Rice University, Houston, Texas, by the US Air Force.

In conjunction with the Rice test, two technologies for recycling surfactants have been demonstrated [45]. As mentioned previously, an economic analysis of the surfactant-enhanced aquifer remediation technology shows that recovery and reuse of the active mixture has a tremendous effect on lowering the cost of the technology [22, 25]. For this reason the AATDF Project funded the study and demonstration of a cost-effective surfactant recovery process. As the contaminants at the Rice site were volatile, it was possible to use air-stripping to regenerate the surfactant. The principal problem with air-stripping, as can well be imagined, is the generation of foam. In addition, the surfactant mixture being used in the Rice system had been optimized to produce a fine, stable foam. Extensive laboratory studies, to be published elsewhere, demonstrated that the surfactant could be air-stripped in a packed tower without foam formation [45]. The key to preventing foaming in the tower proved to be reduction of the liquid loading level. If the liquid loading was raised to the normal design level, which is selected to prevent flooding of the column with the liquid, then foaming could only be avoided by using such a low air rate that mass transfer coefficients were very low. When the liquid loading was reduced sufficiently below that predicted by standard flooding calculations, however, the air rate could be increased substantially without generation of foam. This allowed good mass transfer coefficients to be achieved, along with the resulting effective decontamination of the surfactant.

While packed-column air-stripping is very cost-effective, air-stripping in hollow-fiber membranes is less sensitive to design constraints and consequently is more readily adapted to preventing foam generation. The drawback is that the net cost of hollow-fiber membrane stripping is estimated to be 40% greater than that of packed-column air-stripping. The use of the hollow-fiber air-stripping modules did not completely eliminate operating constraints, however. The initial equipment design for the field test included a positive displacement metering pump, to facilitate determination of flow rates in the unit; however, the pressure variation during the pump cycle was such that foam was generated in the hollow-fiber module. During the low pressure portion of the cycle the air pressure in the module exceeded the liquid pressure by more that a few kilopascals, and air bled into the surfactant solution; during the high pressure stroke, the liquid pressure exceeded the air pressure by more than a few kilopascals, and water bled into the air stream (Figure 2.8). This problem was solved by replacing the constant speed positive displacement metering pump originally installed with a centrifugal pump. The constant pressure produced by the centrifugal pump allowed the relative air and water pressures to be adjusted so that, with the air pressure kept slightly higher than the water pressure, the capillary force inside the porous walls of the hollow-fiber membranes kept the water from leaking through to the air side, thus preventing foam formation. It was also observed, however, that the liquid had to be fed through the hollow-fiber side of the module; when the liquid was fed though the shell side, the pressure drop on the liquid side was too high, so that even when the air stayed on the air side at the top of the module, it would bleed through to the liquid side at the end of the column. By feeding the liquid through the hollow fibers, the pressure drop over the membrane module was kept low and it was possible to prevent both the liquid from bleeding to the air side at the top of the column and the air from leaking to the liquid side at the exit of the module. This may imply that field application of the membrane technology may require parallel air streams and may limit the largest size of membrane modules that can be used.

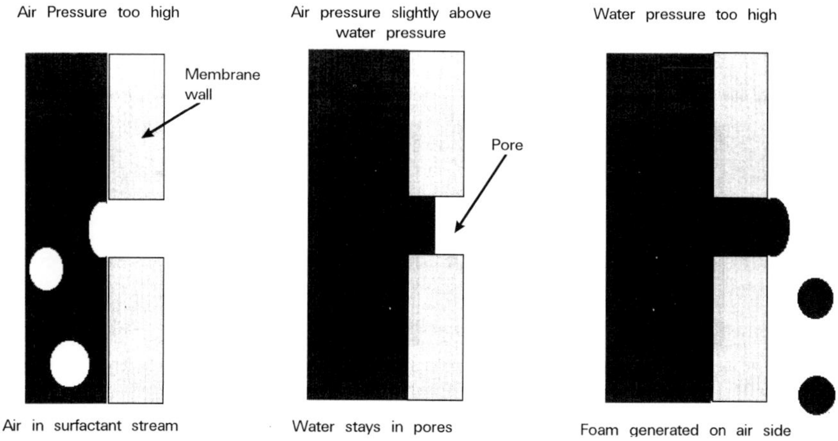

Figure 2.8 Use of porous membranes for air-stripping of NAPL-contaminated surfactant solution.

Besides the surfactant regeneration project just described, the research group at the University of Oklahoma has completed three other field tests of surfactant-enhanced aquifer remediation technology over the last two years [12, 34, 35]. These tests will now be described briefly; full reports of these tests will appear elsewhere. While the technology demonstrated is significant, it is just as significant that the relevant regulatory agencies have shown the flexibility to interpret groundwater protection laws in such a way that the injection of surfactant into the affected aquifers was permitted.

The first of the three field tests was performed in 1995 at Traverse City, Michigan (Table 2.2). To understand the purpose of this field test it is important to realize that during the early stages of the development of surfactant-enhanced aquifer remediation there was great scepticism by many observers that environmental regulators would ever allow injection of surfactants into an active potable water aquifer [36]. It was frequently pointed out that groundwater protection laws strictly prohibit injection of chemicals into ground water. Additionally, the earliest surfactant field tests reported in the refereed literature frequently encountered large unexplained losses of surfactants, which only increased the general scepticism regarding the technology. With these concerns in mind, a field test of surfactant injection was initiated, in 1993, at a US coast guard base in Traverse City, Michigan, on the shore of Lake Michigan. These tests were undertaken with the extensive cooperation of the coast guard, the regional EPA laboratory in Ada, Oklahoma, and the State of Michigan Department of Natural Resources (MDNR). The surfactant was to be injected into a previously contaminated aquifer that had been substantially remediated by innovative bioremediation technologies. The principal goals of this project were to test the surfactant injection with a novel injection technology using a vertical circulation well [35], to demonstrate that by properly matching the surfactant to the aquifer the surfactant could be recovered in quantities approaching 100% [34], and to demonstrate enhanced contaminant removal rates. Probably the most valuable result of the test, however, was to document the procedure necessary to give state and federal regulators responsible for the protection of the affected aquifer confidence that the procedure would not be detrimental to the quality of the ground water. The site turned out to be ideal for this purpose, as the aquifer was both highly conductive and connected to a large freshwater lake at a distance of only 4 km.

Table 2.2 Summary of the Traverse City surfactant injection demonstration

Site background	Contamination	Permit waiver	Field results
US coast guard facility On Lake Michigan Highly conductive (beach sand) Natural flow of 2 m/day Pumping rate of 15 gpm at minimal drawdown	Predominately JP-4 with some PCE Primarily methylated alkanes after 10 years of biotreating Soil cores < 100 mg/kg JP-4, < 1000 mg/kg PCE. PCE groundwater level < 10 ppb	Sand tank tests of well configuration Field tracer test to demonstrate hydraulic control Numerical modeling of potential surfactant losses Minnow/daphnia toxicity < 10 mg/l	Surfactant recovery > 95% PCE removal rate increased by ×40 from 0.6 to 25 µg/l Total hydrocarbon removal rate increased by ×90 from 7 to 620 µg/l

The surfactant chosen for the Traverse City demonstration (Dowfax 8390) was a mixture of mono- and dihexadecyl diphenyloxide disulfonates (DPDS) provided by Dow Chemical Co., Midland, Michigan [18, 19]. Dowfax 8390 meets the US FDA Food Additive Regulation 21 CFR 178.3400 (emulsifiers and/or surface-active agents); it also satisfies US EPA Pesticide Regulation 40 CFR 180.1001 (c) as an inert ingredient in pesticides and is classified as biodegradable based on test criteria set by the Soap and Detergent Association. Earlier studies had shown that the DPDS had a Krafft point below 0°C, was tolerant of calcium concentrations as high as 0.2 M and had adsorptions on common aquifer soils at least an order of magnitude lower than sodium lauryl sulfate. The Traverse City site had evidenced signs of iron precipitation, so precipitation of the DPDS by iron was studied prior to the field test. Ferrous iron (Fe(II)) is the common form of iron in ground water. For 5 mM Fe(II) the surfactant did not precipitate over the range of surfactant concentrations from 0.005 to 50 mM (above and below the approximate critical micelle concentration (CMC) of 6.3 mM). For 0.1 mM Fe(III), the surfactant did not precipitate over the entire range of surfactant concentrations. For 0.5 mM Fe(III) the surfactant precipitated at 0.05 mM. For 1.0 mM Fe(III) the surfactant precipitated over the range from

0.05 to 5.0 mM. While a 5 mM Fe(III) concentration would require a pH not typically found in ground water, the problem illustrates the necessity of matching surfactant behavior to site-specific conditions.

Adsorption of the surfactant on the aquifer materials was determined by measuring the retardation of the surfactant breakthrough curve on injection through soil columns packed with media from the Traverse City site. Two concentrations were used, 0.5 and 5.0 mM. While the surfactant did still adsorb, the amount was substantially less than even other anionics in general. These column studies were repeated with contaminated aquifer media from the Traverse City site to determine typical contaminant removal rates. The principal contaminants at Traverse City were the higher molecular weight and branched alkanes found in the JP-4 jet fuel that had been the original contaminant there; some chlorohydrocarbons were also present, but at only at trace concentrations.

One final laboratory test was required before obtaining a waiver from the required groundwater injection permit. The groundwater remediation treatment system at Traverse City used air-stripping to remove the low levels of dissolved contaminants from the produced water. It was feared that the introduction of the DPDS into the groundwater treatment system would produce foaming that would interfere with the contaminant removal. Foaming tests were completed with the DPDS in the laboratory and it was determined that if the unrecovered concentration of DPDS was kept below the CMC, there was no danger of producing foam in the water treatment system.

The MDNR also requested information regarding the toxicity of the DPDS, which was provided by Dow Chemical Co. Dowfax 8390 has indirect food additive status and has been approved for discharge into wastewater treatment systems at concentrations of 30 mg/l or less. Toxicity tests performed at Dow indicated that Dowfax 8390 was environmentally acceptable in the range below 10 mg/l.

Before the permit waiver was obtained, MDNR requested that laboratory-scale tests of the injection scheme be performed to demonstrate the potential of the injection/extraction configuration to capture essentially all of the injected surfactant. It was observed that by varying the injection and extraction rates this goal could be achieved. MDNR, however, continued to be concerned that the high conductivity of the Traverse City aquifer would allow substantial amounts of the surfactant to escape and eventually reach Lake Michigan, which was only 4 km away, where the aquatic toxicity would be harmful to the animal life in the lake. Consequently, a series of numerical simulations of the Traverse City aquifer was performed, which determined the potential down-gradient concentration profile for the surfactant should some fraction of surfactant be lost from the injection site, rather than being recovered in the water extraction operation. The simulations showed two things. First, by properly adjusting the injection and extraction rates, it should be possible to recover >99% of the surfactant [35, 46]. Secondly, even if 5% of the surfactant was lost, and there was no adsorption of the surfactant on the aquifer materials, the maximum concentration of surfactant at the first monitoring well, located only 3.3 m from the injection wells, would fall below the 10 mg/l toxicity level determined by Dow. Finally, a trial injection of a conservative tracer was tested in a laboratory-scale hydraulic model, in a numerical flow simulator and at the field injection site. There was essentially complete correspondence between the laboratory, numerical simulation and field-scale tracer concentration profiles. This combination of studies was deemed sufficient by the MDNR to allow the field test and they issued a groundwater injection permit waiver. This series of studies provides a framework for developing a standard design procedure for utilization of surfactants in the remediation of aquifers. It especially emphasizes the necessity of performing site-specific studies regarding both the surfactant chemistry and the aquifer hydraulics, including both numerical simulations and field tests, before attempting even a pilot test. While regulatory personnel will always tend to request that 100% recovery of the surfactant be assured, this is impossible in the field. The alternative is to provide regulators with realistic, detailed projections of the fate of the surfactant that escapes the recovery wells.

In the field demonstration 540 gallons of 60 mM (3.6 wt%) Dowfax were injected at a rate of 1 gallon/min into the top screened interval of a single vertical circulation well. After 1 h 40 gallons of water were pumped from the upper interval of the aquitard (the interval in which the surfactant had been injected) to improve overall surfactant recovery. 5 h after the injection of the surfactant, fresh water was pumped into the two nearest down-gradient monitoring wells to form a hydraulic barrier to migration of the surfactant solution out of the flush zone. 20 h after the injection the vertical circulation well was flushed by injecting fresh water into the upper interval. During the entire period the lower interval of the well was pumped at a constant extraction rate. Effluent was sampled to analyze for surfactant, PCE, TCE and selected components of the jet fuel.

Because of the low injection/extraction ratio in the vertical circulation well, the surfactant was diluted below its CMC in the well bore but remained above the CMC through the flushed zone. The injected concentration was 36 g/l, the highest concentration in the flushed zone was 12 g/l and the maximum concentration in the well bore was 3.6 g/l, which was equal to the CMC at aquifer conditions. The principal success of the project was that over 95% of the injected surfactant was recovered over the first 20 h after the injection of the surfactant. The lag in the arrival of the surfactant maximum concentration in the effluent water behind the arrival of a tracer matched the lag observed in the laboratory soil columns. This verified that the adsorption of the surfactant at aquifer conditions was accurately predicted by laboratory adsorption measurements; this is especially important in that adsorption plays an important role in diluting the down-gradient concentration of any surfactant escaping from the treatment zone.

Although the initial contaminant concentrations in the aquifer were already low, there were still significant increases in contaminant removal rates. During the course of the surfactant flush the concentration of PCE in the effluent increased from 5 to 800 μg/l. Two other interesting observations were made, which were unexpected. First, the PCE peak lagged behind the surfactant peak. Since the PCE was solubilized by the surfactant micelles to produce the increased concentrations, it is not obvious how the lag occurred. This phenomenon is observed, however, when surfactant adsorbs to produce admicelles, which subsequently adsolubilize the contaminant. When the water flush then arrives, following the surfactant injection, the desorbed surfactant has a higher chromotographic velocity than the contaminant. The second interesting observation was that the PCE concentration in the effluent remained high (50 μg/l) after the surfactant had been flushed from the system. There is no tested explanation for these phenomena, but it is thought that they may be the result of opening up the low permeability clay materials in the flushed zone so that diffusion of the PCE out of these sources is enhanced even after the surfactant concentration drops below detection limits.

Besides measurement of the surfactant and PCE concentrations, the concentration of total petroleum hydrocarbon (TPH) was also monitored. The TPH concentration in the effluent increased by a factor of 90 to 1 000 μg/l. Over the course of the surfactant flush, over 80% of the total contaminant removal was the direct result of the solubilization of the contaminants by the surfactants.

The Traverse City results, summarized in Table 2.2, emphasize that when the concerns of regulators responsible for protecting the ground water are carefully addressed, it is possible to reach a level of confidence in the process that allowes injection permits to be waived. This requires close cooperation between researchers and regulators. It also requires numerous site-specific studies of the surfactant/aquifer/contaminant interaction. This includes a significant amount of numerical simulation as well as model verification using on-site tracer injection. The role of the surfactant manufacturer is also important, in that Dow provided critical information on aquatic and human toxicity as well as results of standard biodegradation tests. If any of this information had been missing, the test could not have been performed. The Traverse City tests were completed in the summer of 1995.

The research group at the University of Oklahoma completed two additional field tests in summer 1996 [12]. These tests were performed at Hill Air Force Base (HAFB), Ogden, Utah, at designated hazardous waste site Operable Unit 1. They differed dramatically from the Traverse City test for many reasons. Principal among these was that the HAFB site contained a large amount of residual liquid contaminant at the water table. The site had been used both for chemical waste disposal pits, as a hazardous chemical landfill, and for fire training exercises in which jet fuel was pumped into pits and then set on fire so air force firemen could practice putting out jet fuel fires. There are now several decimetres of a liquid solvent/fuel mixture floating on top of the water table and because of the significant variation of the water table seasonally in Ogden, there is significant residual saturation both below and above the water table. The residual liquid is very complex and nearly impossible to characterize completely. The primary constituents are the higher molecular weight and more branched components of the jet fuel, along with dissolved pesticides and chlorinated solvents. There are also polyaromatic compounds and dioxins, probably formed during the partial combustion of the jet fuel. This site is a certified superfund site and remedial action involving installation of a slurry wall into the aquifer to prevent down-gradient migration of dissolved contaminant was ongoing during the test.

Another major difference between the HAFB and Traverse City tests was that the HAFB tests were performed inside sheet pilings. There was concern that solubilization or mobilization of the residual contamination would result in an increase in the down-gradient concentration of pollutants. Consequently, each test cell was separated from the rest of the aquifer by sheet pilings driven 1 m into the clay aquitard at an approximate depth of 10 m. Each cell was tested for leakage and leaking cells were operated in such a way that there was a positive gradient from the aquifer into the cell to keep released contaminant from reaching the aquifer. The use of sheet pilings was expensive but it greatly simplified the waiving of groundwater injection permits and the hydraulics of the actual injection process.

The HAFB tests were part of a joint EPA/DOD/DOE-funded project to compete different advanced remediation technologies against one another in nearly identical settings. Other technologies tested at the same site included injection of 70% alcohol solutions, injection of a tert-butanol/heptanol mixture, injection of cyclodextrane, *in situ* air sparging, vacuum extraction with air injection and *in situ* steam stripping. Only the alcohol injection tests gave remediation results comparable to those obtained with the surfactant injection. The results of these tests were the subject of a symposium at the spring 1997 meeting of the American Chemical Society in San Francisco during April, and of a meeting of the In Situ Flushing Remediation Technology Development Forum in Ogden, Utah, in May 1997.

Because the competing technologies were being compared on as equal a basis as could be achieved in a field setting, the same measures of effectiveness were applied to each test cell. Before installation of the injection wells in the test cells, corings were taken at strategic locations throughout the cells. The corings were then analyzed for the concentrations of selected contaminants. These contaminant distribution results were then analyzed with three-dimensional statistical models of the cells to determine a three-dimensional distribution of the target contaminants throughout the cells. Next, each cell was equilibrated with a constant injection rate of fresh water from the HAFB industrial water supply system. Samples of contaminant concentrations in the effluent waters were taken after equilibration of the cells with the injected water. After equilibration with the injected water, a suite of preselected tracers was added to the injection stream. The tracers consisted of a non-adsorbing tracer like fluorescene or NaBr, plus alcohols like heptanol or 2-hydroxypentanol, which would partition into the residual liquid. The chromatographic movement of the tracers through the contaminated cells was analyzed mathematically to obtain an estimate of the amount of residual NAPL in each cell. In addition, each cell was equipped with a series of multilevel samplers strategically placed throughout the cells. Analysis of the arrival of the tracers at each level of each multilevel sampler allowed a three-dimensional profile of the residual NAPL to be generated. The tracer tests were repeated after the flushings and another set of core samples was obtained and analyzed. Together with an analysis of contaminant concentrations versus time in the effluent from the cells during the flushings, it was possible to determine the effectiveness of the remediation technologies on an essentially comparable basis.

The research group at the University of Oklahoma was responsible for two test cells at HAFB Operable Unit 1. In one cell a solubilization flush was performed, in the other a mobilization flush.

As in the Traverse City tests, extensive laboratory studies were necessary to design the tests and predict the performance of the flushes [34]. Surfactants were screened on the basis of their phase behavior and adsorption in the aquifer materials (soil and water, including the proper electrolyte composition) at the aquifer temperature. They were also tested in the HAFB water that would be injected into the cells. The same Dow DPDS (Dowfax 8390) was used in the solubilization cell as was used at Traverse City.

For the mobilization test, a new technique was developed for designing the surfactant mixture that was to be used to create the microemulsion *in situ*. The procedure generally used in EOR for formulating a single-phase microemulsion is to vary the electrolyte composition for a single surfactant species; when the surfactant begins to form liquid crystals, an alcohol is introduced at sufficient concentration to dissolve the liquid crystals. This is the so-called salinity scan (Figures 2.9 and 2.10) technique originally developed by Winsor [47]. In EOR applications, the use of an alcohol is undesirable because of the likelihood of chromatographic separation between the alcohol and the surfactant during the propagation of the low pore volume surfactant slug through the reservoir [23, 38]. Alcohol was eliminated from EOR formulations by introducing branching or ethoxy groups into the surfactant between the headgroup and the hydrophobe, thus breaking up the liquid crystals without utilization of the alcohol. In aquifer remediation, however, the close well spacings on smaller volumes of the aquifers versus the reservoirs of EOR allow multiple pore volume surfactant slugs to be used. This minimizes the effects of chromatographic separation of the slug components and allows formulations to be based on multicomponent surfactant mixtures. Thus, our formulations of Winsor Type III systems with the Hill NAPL were based on mixtures of water-soluble and oil-soluble surfactants (Figure 2.11) at a fixed electrolyte concentration, without the presence of a hydrotrope or alcohol (called co-surfactants in the microemulsion literature).

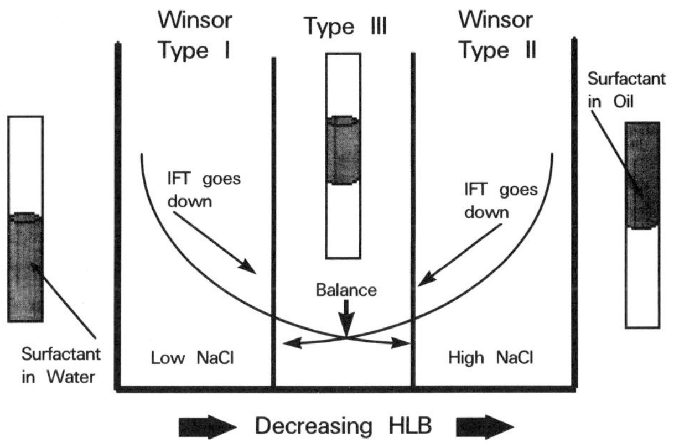

Figure 2.9 Salinity scan for producing ultra-low interfacial tension between NAPL and ground water.

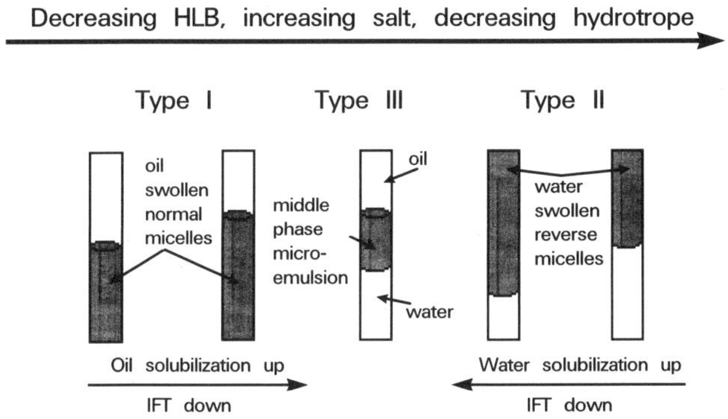

Figure 2.10 Correlation between decreasing interfacial tension and increasing solubilization.

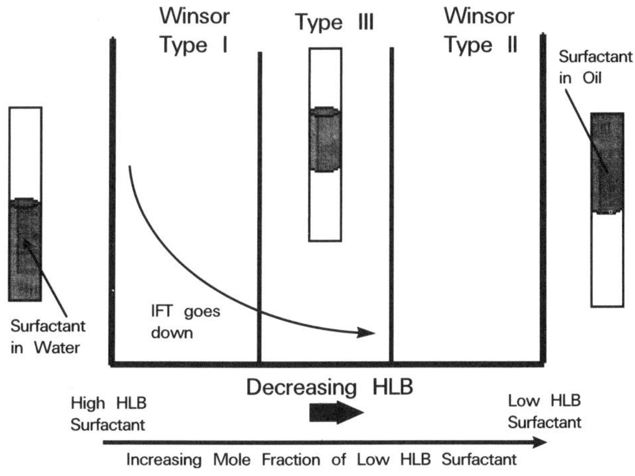

Figure 2.11 Location of low interfacial tension composition by blending of a water-soluble surfactant and an oil-soluble surfactant.

It was discovered, however, that variation of the weight ratio of two dissimilar surfactants sometimes does not give sufficient control of the phase behavior to optimize the interfacial tension lowering. Hence the introduction of a hydrotrope scan (Figure 2.12) to find the optimum formulation for each surfactant mixture studied. In particular, it was found that varying concentrations of sodium mono- and dimethyl naphthalene sulfonate produced optimum phase behavior. Thus, whereas traditional EOR middle-phase formulations include substantial amounts of alcohol and are characterized by the finding of an optimal salinity, it was demonstrated that a formulation for mobilization is feasible based on an optimal hydrotrope concentration and containing no alcohol. However, some traditional, alcohol-containing systems were also formulated and the so-called optimum salinities of these systems versus the Hill NAPL were found. In subsequent column tests it was found that the new style formulations gave NAPL recoveries equal to those obtained with the alcohol-containing formulations, but they had the additional advantage of not including any volatile components. By use of the new formulations interfacial tensions were obtained down to 10^{-3} mN/m and solubilizations of over 500 000 mg/l were found. At solubilizations of 500 000 mg/l the middle phase was not clear but was quite cloudy, with an optical density similar to skimmed milk; analysis of the middle phase did not give any evidence of the presence of liquid crystals, despite the cloudiness. The behavior of the interfacial tension versus hydrotrope concentration over the range of the HLB or hydrotrope scans was identical to that expected in traditional Winsor-type salinity scans with alcohols.

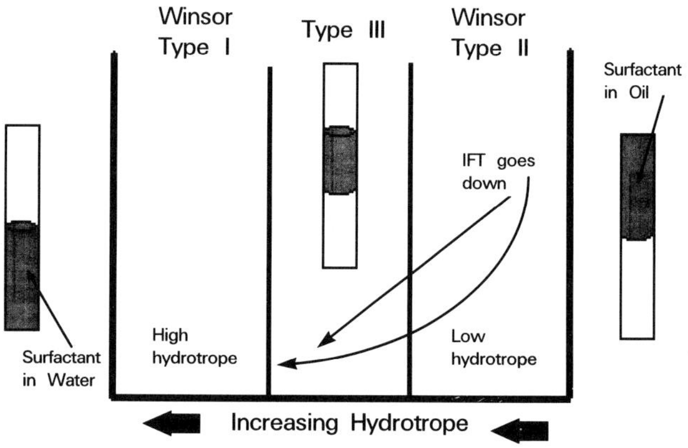

Figure 2.12 Use of varying hydrotrope concentration to locate low interfacial tension system.

Both the solubilization system and the mobilization system were tested in soil columns before the field tests. The soil column tests were performed with soil from the test cells, with water from the test cells and with samples of the Hill

NAPL used to create the residual saturation in the soil columns. The final formulation for the solubilization cell was a 4.3 wt% solution of a Dow DPDS, Dowfax 8390. The mobilization formulation was 2.2 wt% of a di(2-ethylhexyl) sulfosuccinate sodium salt (Cytec AOT), 2.1 wt% of an ethoxylated sorbitol fatty acid ester of HLB 15.8 (ICI Tween 80), with $CaCl_2$ used to obtain the optimum phase behavior. In the soil column tests 3.8 pore volumes of the mobilization formulation removed 58% of the residual NAPL with 90% recovery of the surfactant; the solubilization formulation gave 44% removal of the NAPL with 98% recovery of the surfactant after 12 pore volumes (Table 2.3).

Table 2.3 Soil column results for Hill laboratory studies

Surfactant	Sn_i (%)	Sn_f (%)	Removal (%)	Surfactant recovery (%)
AOT/Tween/$CaCl_2$ 3.8 PV	8.4	4.8	58	90
Dowfax 8390 12 PV	8.4	4.7	44	98

Sn_i = initial saturation with Hill NAPL.
Sn_f = final saturation with Hill NAPL.
PV = pore volume.

The analysis of the results of the field test was not complete at the time of writing. Preliminary results, however, show that the field tests performed according to design. After injection of 10 pore volumes of the 4.3 wt% DPDS solution, >95% of the surfactant was recovered and tracer tests indicated removal of 45% of the NAPL originally in the cell. Injection of 6.6 pore volumes of the mobilization formulation resulted in removal of 74% of the original NAPL, along with recovery of >90% of the surfactant (Table 2.4). Analysis of post-flush soil cores from the mobilization cell for the target contaminants indicated an average removal of 91% of the target contaminants (Table 2.5).

Table 2.4 Partitioning tracer results for Hill field tests

Surfactant used	Well no.	Initial NAPL saturation	Final NAPL saturation	Fraction of NAPL removed
Dowfax 8390[a]	EW 1	0.043	0.028	35
solubilization (4.3%,	EW 2	0.083	0.040	52
10 PV)	EW 3	0.084	0.052	32
AOT/Tween 80	EW 1	0.086	0.020	77
mobilization	EW 2	0.072	0.014	81
(2.2%/2.1%, 6.6 PV)	EW 3	0.103	0.037	64

[a] Dowfax 8390 recovery > 95%.
PV = pore volume.

To this date there is no agreed explanation for the apparent discrepancy between the >90% removal indicated by the soil cores and the 74% removal indicated by the tracer tests. An almost identical discrepancy was observed in the alcohol flushes. One explanation that has been suggested is that the surfactant and alcohol flushings resulted in the separation of the insoluble materials like PAHs and asphaltenes in the Hill NAPL, which were then deposited in the cell on the soil and appeared to the partitioning tracers as residual liquids. This explanation is not consistent, however, with the observation of middle-phase formation in the laboratory studies, where no separation of the NAPL into components was observed. A possible explanation is that asphaltene-type materials were present, but that they were firmly adsorbed onto the soils and consequently were left behind when the NAPL was mobilized; thus, they would not be expected to be present in the samples of NAPL obtained from the HAFB site and used in the phase behavior studies.

Table 2.5 Hill contaminant removal based on pre-flush and post-flush soil cores

Contaminant	Initial (mg/l)	Final (mg/l)	Removal (%)
Toluene	2.04	0.05	97.6
o-Xylene	2.65	0.08	96.8
Trimethylbenzene	2.53	0.18	92.8
Decane	30	2.75	90.8
DCB	4.48	1.11	75.2
Naphthalene	2.25	0.17	92.4

2.4 Summary and conclusions

The potential for using surfactants for rapid, relatively economical remediation of contaminated aquifers has been obvious since the days of the excitement of EOR. There has been a general reluctance, however, to use a sophisticated chemically based technology to undo the environmental damage done by the accidental release or improper disposal of man-made chemicals. With the passing of time the environmental community has come to realize that the extent and depth of damage done by the widespread release of hydrocarbon and chlorocarbon liquids into ground water requires drastic measures if significant progress is to be made in a reasonable period of time and with a reasonable amount of money. Since the door was opened to the use of chemical enhancements to remediation projects in the early 1990s there has been an almost explosive growth in development of surfactant-enhanced aquifer remediation. The technology was able to grow rapidly by borrowing extensively from earlier work on EOR. Now that the differences and similarities between aquifer remediation and EOR are becoming better delineated, and as the opportunity for field tests grows, full-scale applications to major contaminated sites are only a few years away.

The development of surfactant-enhanced aquifer remediation as an important new application of surfactants opens up a number of new areas for research and development. Foremost among these is the need to study the new types of microemulsion systems that are being developed. The use of surfactant mixtures and the use of hydrotropes to formulate microemulsions goes beyond the extensive microemulsion literature that has developed over the last 20 years. There is also a dearth of robust surfactants capable of forming microemulsions with DNAPLs, showing high hardness tolerance and low adsorption, and being readily treated for contaminant removal. The design of an ideal surfactant for all of these requirements has not been undertaken anywhere, to the authors' knowledge. Furthermore, it would be desirable to develop a suite of surfactant isomers that could be blended to produce the desired phase behavior under different conditions of DNAPL or LNAPL composition, and different aquifer temperature and electrolyte conditions.

Another critical area of study is the natural biodegradation of surfactants in aquifers. This knowledge is critical both for assessing the long-term impact of residual surfactant left in the treated zone of a contaminated aquifer and for analyzing the risk associated with surfactant that escapes from a surfactant injection site and is released into the less contaminated parts of an aquifer. What is the potential impact, for example, if micelles of a surfactant containing solubilized contaminants escape the confinement zone and begin to migrate toward a drinking-water supply, a lake or a stream? Not enough is yet known about surfactant biodegradation in aquifers or the interaction of adsorption with biodegradation and migration to give a confident answer to this question [37]. At many potential sites for surfactant-enhanced aquifer remediation this question may well be critical in determining if surfactant technology can be used.

Acknowledgements

Partial financial support for this work was provided by the US Environmental Protection Agency through its Office of Research and Development. It has not been subjected to Agency review and therefore does not necessarily reflect the views of the Agency, and no official endorsement should be inferred.

Support was also provided by the Oklahoma Center for the Advancement of Science and Technology (OCAST) and the sponsors of the Institute for Applied Surfactant Research (IASR) at the University of Oklahoma, including Akzo Nobel, Amway, Colgate-Palmolive, Dow Chemical Co., DowElanco, DuPont, Henkel, ICI, Kerr-McGee, Lever, Lubrizol, Nikko Chemical, Phillips Petroleum, Pilot Chemical, Reckitt and Coleman, Shell, Sun and Witco.

Dr Harwell holds the Conoco-DuPont Professorship in Chemical Engineering at the University of Oklahoma.

When this chapter was written, Dr Sabatini was on sabbatical from the School of Civil Engineering and Environmental Science, University of Oklahoma, as a Senior Fulbright Scholar at the Lehrstuhl fuer Angewandte Geologie, Universität Tuebingen, Germany.

Surfactant used in the tests at Traverse City and HAFB was donated by Dow Chemical Co. and ICI.

References

[1] Haley, J.L., Hanson. B., Enfield, C., and Glass, J. (1991) Evaluating the effectiveness of ground water extraction systems. *Ground Water Monitoring Review*, Winter, 119-24.

[2] Palmer, C.D. and Fish, W. (1992) *Chemical Enhancements to Pump and Treat Remediation*, US, Environmental Protection Agency, EPA 540-S-92-001, Ada, Oklahoma.

[3] Sabatini, D.A., Harwell, J.H., Deshpande, S., Wesson, L. and Wade, D. (1997) Dowfax surfactant components in surfactant-enhanced soil remediation. *Water Research* (in press).

[4] Nayyar, S.P., Sabatini, D.A. and Harwell, J.H. (1994) Surfactant adsolubilization and modified admicellar sorption of nonpolar, polar, and ionizable organic contaminants. *Environmental Science and Technology*, **28** 1874-81.

[5] Bowman, R.S., Haggerty, G.M., Huddleston, D.N and Flynn, M.M. (1995) in *Surfactant-Enhanced Subsurface Remediation: Emerging Technologies* (eds D.A. Sabatini, R.C. Knox and J.H. Harwell), ACS Symposium Series 594, American Chemical Society, Washington, DC, pp. 54-64.

[6] Morrow, N.R., Chatzis, I. and Taber, J.J. (1988) Entrapment and mobilization of residual oil in bead packs. *SPE Reservoir Engineering*, **3** (3) 927-34.

[7] Federal Facilities Policy Group (1995) *Improving Federal Facilities Cleanup: Report of the Federal Facilities Policy Group*, Council on Environmental Quality, Office of Management and Budget, Washington, DC, http://keystone.org/spp/publicat.html.

[8] EPA (1995) *The Annual Status Report on Innovative Treatment Technologies*, The Innovative Treatment Database. US Environmental Protection Agency, EPA 542-R-95-008, Washington, DC, http://clu-in.com/asrl.htm.

[9] Sabatini, D.A., Knox, R.C. and Harwell, J.H. (1996) *Surfactant-Enhanced DNAPL Remediation: Surfactant Selection, Hydraulic Efficiency, and Economic Factors*, US Environmental Proection Agency Environmental Research Brief, EPA/600/S-96/002, Ada, Oklahoma.

[10] Sabatini, D.A., Knox, R.C. and Harwell, J.H. (eds) (1995) *Surfactant Enhanced Subsurface Remediation: Emerging Technologies*, ACS Symposium Series 594, American Chemical Society, Washington, DC, pp. 312.

[11] West, C.C. and Harwell, J.H. (1992) Surfactants and subsurface remediation. *Environmental Science and Technology*, **26** 2324-30.

[12] Sabatini, D.A., Harwell, J.H. and Knox, R.C (1997b) *Surfactant Selection Criteria for Enhanced Subsurface Remediation*, ACS Symposium Series, 213th American Chemical Society National Meeting and Exposition, San Francisco, CA., April 13–17, American Chemical Society, Washington, DC (in press).

[13] Vignon, B.W. and Rubin, A.J. (1989) *Journal of the Water Pollution Control Federation*, **61** (7) 1233-40.

[14] Abdul, A.S., Gibson, T.L., Ang, C.C., Smith, J.C. and Sobczynski, R.E. (1992) Selection of surfactants for the removal of petroleum products from shallow sandy aquifers. *Ground Water*, **30** (2) 219-31

[15] Valsaraj, K.T. and Thibodeaux, L.J. (1989) Relationship between micelle-water and octanol-water partition constants for hydrophobic organics of environmental interest. *Water Research*, **23** (2) 183-89.
[16] Edwards, D.A., Luthy, R.G. and Liu, Z. (1991) Solubilization of polycyclic hydrocarbons in micellar non-ionic surfactants solutions. *Environmental Science and Technology*, **25** (1) 127-33.
[17] West, C.C. (1992) in *Transport and Remediation of Subsurface Contaminants: Colloidal, Interfacial and Surfactant Phenomena* (eds D.A. Sabatini and R.C. Knox), ACS Symposium Series 491, American Chemical Society, Washington, DC, pp. 149-58.
[18] Rouse, J.D., Sabatini, D.A. and Harwell, J.H. (1993) Minimizing surfactant losses using twin-head anionic surfactants in subsurface remediation. *Environmental Science and Technology*, **27** (10) 2072-78.
[19] Rouse, J.D., Sabatini, D.A., Deeds, N.E., Brown, E. and Harwell, J.H. (1995) Influence of anionic surfactants on bioremediation of hydrocarbons. *Environmental Science and Technology*, **29** (10), 2484-89.
[20] Shiau, B.J., Sabatini, D.A. and Harwell, J.H. (1994) Solubilization and microemulsification of DNAPLs using direct food additive (edible) surfactants *Ground Water*, **32** (4) 561-69.
[21] Pennel, K.D., Abriola, L.M. and Weber, W.J., Jr. (1993) Surfactant enhanced remediation of soil columns contaminated by residual tetrachloroethylene. *Environmental Science and Technology*, **27** (12) 2341-51.
[22] Krebs-Yuill, B., Harwell, J.H., Sabatini, D.A. and Knox, R.C. (1995) Economic considerations in surfactant-enhanced pump-and-treat, in *Surfactant-Enhanced Subsurface Remediation: Emerging Technologies* (eds D.A. Sabatini, R.C. Knox and J.H. Harwell), ACS Symposium Series 594, American Chemical Society, Washington, DC, 265-79.
[23] Shah, D.O. and Schechter, R.S. (eds.) (1977) *Improved Oil Recovery by Surfactant and Polymer Flooding*, Academic Press.
[24] Pope, G.A. and Wade, W.H. (1995) Lessons from enhanced oil recovery research for surfactant-enhanced aquifer remediation, in *Surfactant-Enhanced Subsurface Remediation: Emerging Technologies* (eds D.A. Sabatini, R.C. Knox. and J.H Harwell), ACS Symposium Series 594, American Chemical Society, Washington, DC, pp. 142-60.
[25] Krebs-Yuill, B., Harwell, J.H., Sabatini, D.A., Quinton, G.E. and. Shoemaker, S.H (1996) Economic Study of Surfactant-Enhanced Pump-and-Treat Remediation, 69th Annual Water Environment Federation Conference, Dallas, Texas, October 5–9 (in preparation).
[26] Keely, J.F. (1989) *Performance Evaluation of Pump-and-Treat Remediations*, US Environmental Protection Agency, EPA 540-89-005, Ada, Oklahoma.
[27] Stallard, M. and Anderson, E. (1992) *Aquifer Restoration: Pump-and-Treat and the Alternatives*, AGWSE Educational Program, Las Vegas, NV.
[28] Armstrong, J.E., Frind, E.O. and McClellan, R.D. (1994) Nonequilibrium mass transfer between the vapor, aqueous and solid phases in unsaturated soils during vapor extraction. *Water Resources Research*, **30** (2) 355-68.
[29] Voudrias, E.A., and Yeh, M. (1994) Dissolution of a toluene pool under constant and variable hydraulic gradients with implications for aquifer remediation. *Ground Water*, **32** (2), 305-11.
[30] Shiau, B.J., Sabatini, D.A. and Harwell, J.H. (1995) Properties of food grade surfactants affecting subsurface remediation of chlorinated solvents. *Environmental Science and Technology*, **29** (12) 2929-35.
[31] Shiau, B.J., Sabatini, D.A., Harwell, J.H. and Vu, D. (1996) Microemulsions of mixed chlorinated solvents using food grade (edible) surfactants. *Environmental Science and Technology*, **30** (1) 97-103.
[32] Shiau, B.J., Sabatini, D.A. and Harwell, J.H. (1997) Removal of chlorinated solvents in subsurface media using edible surfactants: column studies. *Journal of Environmental Engineering Division—ASCE* (accepted).

[33] Rouse, J.D., Sabatini, D.A., Brown, R.E. and Harwell, J.H. (1996) Evaluation of ethoxylated alkylsulfate surfactants for use in subsurface remediation. *Water Environment Research*, **68** (2) 162-68.

[34] Sabatini, D.A., Knox, R.C., Harwell, J.H., Soerens, T., Chen, L., Brown, R.E. and West, C. (1997) Design of a surfactant remediation field demonstration based on laboratory and modeling studies. *Ground Water* **35** (6) 948-53.

[35] Knox, R.C., Sabatini, D.A., Harwell, J.H., Brown, R.E., West, C.C., Blaha, F. and Griffin, S. (1997) Surfactant remediation field demonstration using a vertical circulation well. *Ground Water* **35** (6) 954-63.

[36] Steimle, R. and Kovalick, W.W. (1996) 'State Policies Concerning the Use of Injectants for in Situ Ground Water Remediation, EPA 542-R-96-001, http://www.gnet.org/forsite.

[37] Rouse, J.D., Sabatini, D.A., Sulflita, J.M. and Harwell, J.H. (1994) Influence of surfactants on microbial degradation of organic compounds. *Critical Reviews in Environmental Science and Technology*, **24** (4) 325-70.

[38] Bourrel, M. and Schechter, R.S. (1988) *Microemulsions and Related Systems*, Surfactant Science Series, vol. 30, Marcel Dekker, Inc., New York.

[39] Wu, B., Shiau, B.J., Sabatini, D.A., Vu, D.Q. and Harwell, J.H. (1997) Formulating microemulsion systems of petroleum hydrocarbons using surfactant–cosurfactant mixtures. *Ground Water* (submitted).

[40] Lipe, M., Sabatini, D.A., Hasegawa, M.A., and Harwell, J.H. (1996) Micellar enhanced ultrafiltration and air stripping for surfactant-contaminated separation and re-use. *Ground Water Monitoring and Remediation*, **16** (1) 85-92.

[41] Hasegawa, M.A., Sabatini, D.A. and Harwell, J.H. (1997) Liquid-liquid extraction for surfactant-contaminant separation and surfactant reuse. *Journal of Environmental Engineering*, **123** (7) 691-97.

[42] Fountain, J.C. (1992) A comparison of field tests of surfactant flooding: examples of mobility control of DNAPL, in *Transport and Remediation of Subsurface Contaminants: Colloidal, Interfacial and Surfactant Phenomena* (eds D.A. Sabatini and R.C. Knox), ACS Symposium Series 491, American Chemical Society, Washington, DC, 182-91.

[43] Fountain, J.C., Waddell-Sheets, C., Lagowski, A., Taylor, C., Frazier, D. and Byrne, M. (1995) Enhanced removal of dense nonaqueous-phase liquids using surfactants: capabilities and limitations from field trials, in *Surfactant-Enhanced Subsurface Remediation: Emerging Technologies* (eds D.A. Sabatini, R.C. Knox and J.H. Harwell), ACS Symposium Series 594, American Chemical Society, Washington, DC, 177-90.

[44] Brown, C.L., Delshad, M., Dwarakanath, V., Jackson, R.E., Londergan, J.T., Meinardus, H.W., McKinney, D.C., Oolman, T., Pope, G.A. and Wade, W.H. (1997) A successful demonstration of surfactant flooding of an alluvial aquifer contaminated with DNAPL. *Environmental Science and Technology* (in press).

[45] Walk, R.D. (1997) Design and scale-up of a packed column and hollow fiber membrane air stripping system, M.Sc. Thesis, The University of Oklahoma, Norman, Oklahoma.

[46] Chen, Y., Chen, L.Y. and Knox, R.C. (1995) Modeling the effectiveness of innovative measures for improving the hydraulic efficiency of surfactant injection and recovery, in *Surfactant-Enhanced Subsurface Remediation: Emerging Technologies* (eds D.A. Sabatini, R.C. Knox and J.H. Harwell), ACS Symposium Series 594, American Chemical Society, Washington, DC, pp. 249-64.

[47] Winsor, P.A. (1968) *Chem. Rev.*, **68** (1).

3 Current developments in cationic surfactants

Dale S. Steichen and J. Fred Gadberry

3.1 Introduction and background

Cationic surfactants represent one of the smaller classes of surfactants when compared to anionic and non-ionic surfactants. An annual volume estimate for world-wide production of cationics is 700 000 tonnes [1]. However, the uniqueness of the positively charged hydrophile provides specific properties that for many applications makes these materials indispensable. The surfactants in the class are dominated by a positively charged nitrogen as the core hydrophile. While other positively charged hydrophiles are possible, such as sulfonium and phosphonium, virtually no commercial products of this type exist [2].

Nitrogen-based cationics were developed by the Armour company in the 1940s as a means to utilize the tallow from their Chicago stockyards [3]. The first significant product was dihydrogenated tallow dimethyl ammonium chloride, which today still finds use in a variety of applications where a positive charge and hydrophobation are required. The single largest market for cationic surfactants is as the active ingredient in fabric softeners [4]. The market for fabric softeners was established as a consequence of the move to the higher performing synthetic anionic detergents, such as branched linear alkyl benzene sulfonates [5]. The trade-off for the high performing surfactants was fabrics that possessed an unacceptably rough handfeel. Fabric softeners made from dihydrogenated tallow dimethyl ammonium chloride compensated for this, allowing the consumer to have clean and soft clothes. A second significant market for cationics is their use as the hydrophobation agent in organoclays. The US market for organoclays is approximately 40 million lbs (Klein and Co., private communication).

3.2 Manufacturing processes

3.2.1 Amine preparation

The manufacturing process for cationic surfactants can be divided into two parts. The first part is the creation of an alkylated amine. Several processes can achieve this end-point and they are briefly reviewed below. The largest volume process, which was developed by Armour, starts with tallow triglyceride, which is split to yield fatty acid and glycerine. The fatty acid is reacted with ammonia and converted to fatty nitrile under high pressure and temperature conditions [3]. The nitrile is then hydrogenated to a primary amine from which amine derivatives can be made using a variety of reagents [6, 7]. The example in equations (3.1) to (3.3) shows the conversion of fatty acid to dialkyl secondary amine via catalytic deammoniafication. (For this and all subsequent structures $n = 8–20$ unless otherwise noted.)

$$CH_3(CH_2)_nC(=O)OH + NH_3 \xrightarrow[\text{heat}]{\text{cat.}} CH_3(CH_2)_nC{\equiv}N + H_2O \tag{3.1}$$

$$CH_3(CH_2)_nC{\equiv}N \xrightarrow[H_2]{\text{cat.}} CH_3(CH_2)_nCH_2NH_2 \tag{3.2}$$

$$CH_3(CH_2)_nCH_2NH_2 \xrightarrow[-NH_3]{\text{cat./}H_2} (CH_3(CH_2)_nCH_2)_2NH \tag{3.3}$$

Alternative routes to amine derivatives have been developed in the intervening years. The production of dimethylamines can be accomplished by the routes shown in equations (3.4) to (3.7) [8]. Both routes involve the reaction of an alkyl halide with dimethyl amine. The first route is the conversion of a fatty alcohol to fatty chloride using phosphorous trichloride. The alkyl chloride is reacted with dimethylamine giving the alkyl dimethylamine [9, 10]:

$$CH_3(CH_2)_nOH + PCl_3 \longrightarrow CH_3(CH_2)_nCl + H_3PO_3 \tag{3.4}$$

$$CH_3(CH_2)_nCl + (CH_3)_2NH \longrightarrow CH_3(CH_2)_nN(CH_3)_2 + (CH_3)_2NH_2^+\ Cl^- \tag{3.5}$$

The second route reacts an alpha olefin derived from ethylene with hydrogen bromide and a free radical initiator, resulting in an alkyl bromide. The alkyl bromide is reacted with dimethylamine providing the desired product [11, 12]:

$$H_2C{=}CH(CH_2)_nCH_3 + HBr \longrightarrow BrCH_2CH_2(CH_2)_nCH_3 \tag{3.6}$$

$$BrCH_2CH_2(CH_2)_nCH_3 + 2(CH_3)_2NH \longrightarrow (CH_3)_2NCH_2CH_2(CH_2)_nCH_3 + (CH_3)_2NHBr \tag{3.7}$$

Alcohols and aldehydes are also suitable materials for the creation of an alkyl amine. In addition to the aforementioned formation of alkyl chloride as an intermediate, alcohols can be directly converted to amines under hydrogenation conditions in the presence of ammonia while aldehydes are pre-reacted to form the imine followed by hydrogenation [13]. Selectivity to the primary amine with these techniques is difficult and this process is more typically used for the preparation of tertiary amines, where the reaction can be driven to completion. In certain cases alcohols and aldehydes provide structural elements that are not attainable from natural sources. An example is the formation of a hydrogenated tallow 2-ethyl hexyl methyl amine. The amine is prepared as shown in equation (3.8) using a hydrogenated tallow amine reacted with 2-ethyl hexanal [14, 15]:

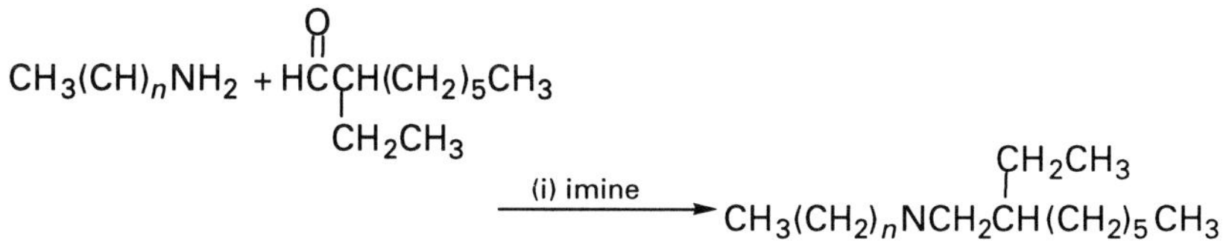

(3.8)

The intermediate imine is hydrogenated, giving the secondary amine. Formation of tertiary imine is suppressed due to the steric hindrance of the branched-chain substituent. The cationic derived from this amine is highly water soluble due to suppression of lamellar phase formation by the branched alkyl group.

Nitrogen cationic surfactants can also be created by the use of difunctional small molecule amines, which, after formation of an amide or ester bond, leave an amine residue that is suitable for quaternization, as shown in equations (3.9) to (3.11). The amine residue is then reacted with a suitable alkylating agent to form the cationic. Similarly, reaction of a triglyceride with diethylene triamine gives initially the diamide, which under appropriate conditions can be cyclized to the imidazoline [16].

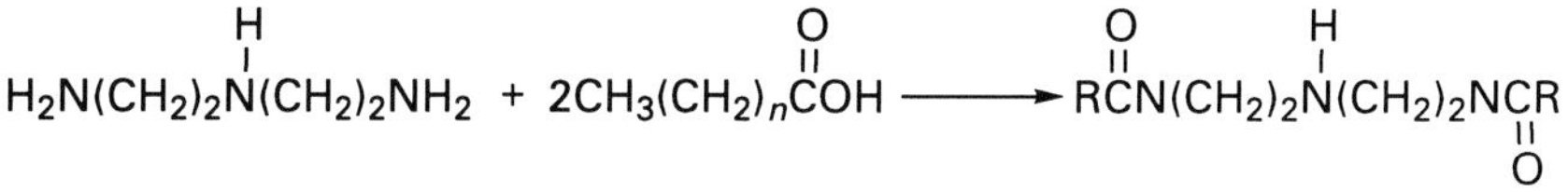

(3.9)

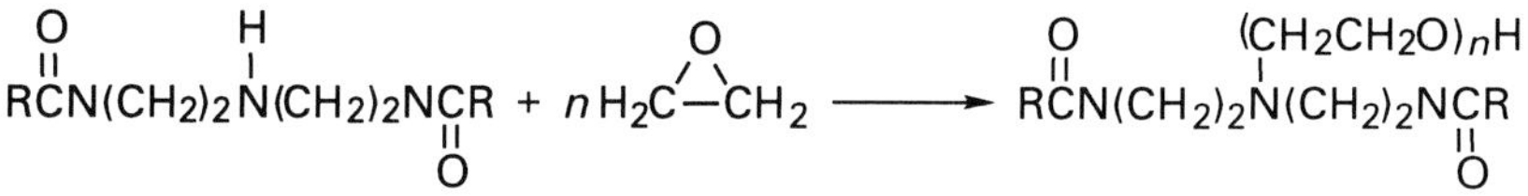

(3.10)

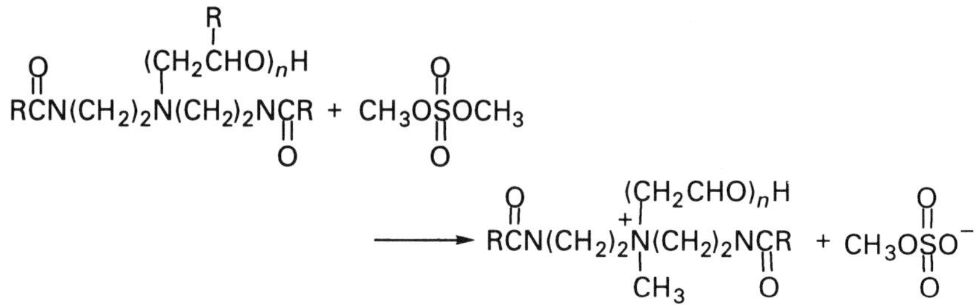

(3.11)

3.2.2 *Quaternization*

Cationic surfactants are prepared by the reaction of a tertiary amine with an alkylating agent, resulting in the formation of the positively charged nitrogen center. This reaction can be accomplished with a variety of agents, as shown in Table 3.1 where each R represents a hydrocarbon residue.

Table 3.1 Typical quaternization agents

Reagent	CAS number	Product
Dimethyl sulfate	77-78-1	$RR'R''N^{+}CH_3\ ^{-}OSO_2OCH_3$
Methyl chloride	74-87-3	$RR'R''N^{+}CH_3\ Cl^{-}$
Benzyl chloride	100-44-7	$RR'R''N^{+}CH_2C_6H_5\ Cl^{-}$
Ethyl chloride	75-00-3	$RR'R''N^{+}CH_2CH_3\ Cl^{-}$
Alkyl chloride	Various	$RR'R''N^{+}R'''\ Cl^{-}$

The most commonly used alkylating agents are dimethyl sulfate and methyl chloride [16]. These materials find common use due to their widespread availability and modest cost structure. Each of these materials is considered to be toxic and dimethyl sulfate is classified as a cancer suspect agent (Registry of Toxic Effects of Chemical Substances, RTECS# WS8225000) The material safety data sheet (MSDS) and manufacturer's information must be carefully reviewed prior to laboratory or larger scale preparations using these materials. Methyl chloride is typically reacted with a suitable amine at 90 to 120°C in a pressure vessel rated to 200 psi [16]. The methyl chloride is added over the course of the reaction, often on a pressure demand basis. Excess methyl chloride is added to drive the reaction to a low free amine content of 1 or 2%. The excess methyl chloride is then vented from the vessel and scavenged with an appropriate scrubber. Sparging with an inert gas can further lower the final methyl chloride level to lower than 1000 ppm.

Dimethyl sulfate is usually reacted with an amine at 40 to 60°C in a closed vessel but this can be done at higher temperatures to increase the fluidity of the reaction medium [17]. The reaction can be completed under atmospheric conditions. For safety, a slightly negative pressure can be maintained to ensure that any dimethyl sulfate vapors are contained within the vessel. The reaction is exothermic, yielding 141 kJ/kg in the case of the reaction of dimethyl sulfate with a triethanol amine tallow fatty acid ester [18]. As a consequence of this exothermicity it is imperative to add the dimethyl sulfate slowly over the course of the reaction. For large-scale preparations sufficient cooling must be available on the reactor to maintain temperature control and adequate agitation must be provided to prevent 'hot spots'. The dimethyl sulfate is added sub-stoichiometrically to ensure that free amine is available at the end of the reaction. The presence of free amine ensures that all of the dimethyl sulfate has been reacted.

Most quaternizations are accomplished in the presence of a solvent because of the high melting points of most cationic surfactants. The solvent can range in concentration from 85% to as low as 10%, and in some cases the quaternizations can be run neat [17]. Alcohols are the solvents of choice due to their hydrotropic properties. The most popular choices include ethanol and 2-propanol. Glycols are increasing in popularity due to concerns over volatile organic compounds (VOCs). Water is a viable solvent for many cationic surfactants but only in a few cases can the concentration of the cationic surfactant exceed 40% [18].

3.3 Applications of cationic surfactants

3.3.1 Fabric softeners

Fabric softeners represent the single largest outlet for cationic surfactants, consuming 200 000 tonnes in 1992 [19]. The basic structural requirement for a surfactant useful for fabric softening is the presence of two alkyl groups, each with 12 to 18 carbon atoms, and at least one positively charged hydrophile, which in commercial fabric softeners is exclusively nitrogen (Figure 3.1).

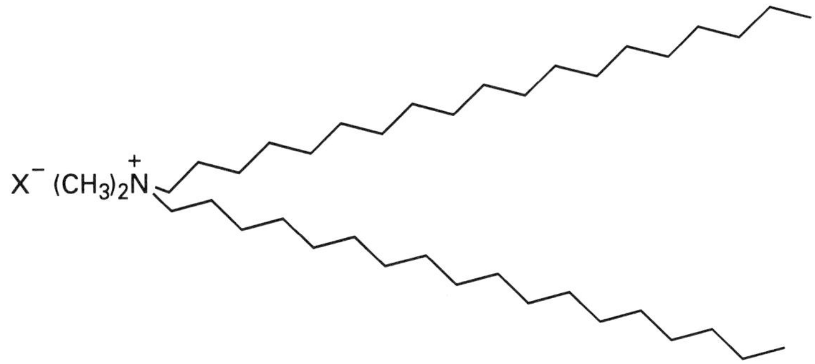

Figure 3.1 Basic structure of a surfactant.

The positively charged nitrogen aids in the deposition of the surfactant onto the fabric and the 'twin tails' provide the lubricity and handfeel desired by consumers [20]. From this basic structure a medley of materials arise, each with their particular exigency being satisfied.

3.3.1.1 Structures. The driving force for the development and existence of the structural forest of cationic softeners is the biodegradability requirement of the EU [5]. This directive, which requires a minimum of 60% biodegradability based on a 28-day closed-bottle test with non-acclimated bacteria, doomed the use of nitrile-based quaternaries in many countries within Europe. Subsequent studies on the environmental fate of nitrile-based quaternaries has questioned whether the closed-bottle test is an accurate assessment of ultimate environmental fate [21, 22]. Nevertheless, the requirements stand, driving a new generation of cationic softeners based on structures with enhanced biodegradability.

The new structures place a cleavable functionality within the 'twin tails' of the cationic surfactant. The cleavable functionality aids in biodegradation by separating the positively charged nitrogen hydrophile from the hydrophobic portion, as shown in Figure 3.2. The hydrophobic portion is typically a fatty alcohol or fatty acid, both of which are materials found in nature that exhibit excellent biodegradability [23–29].

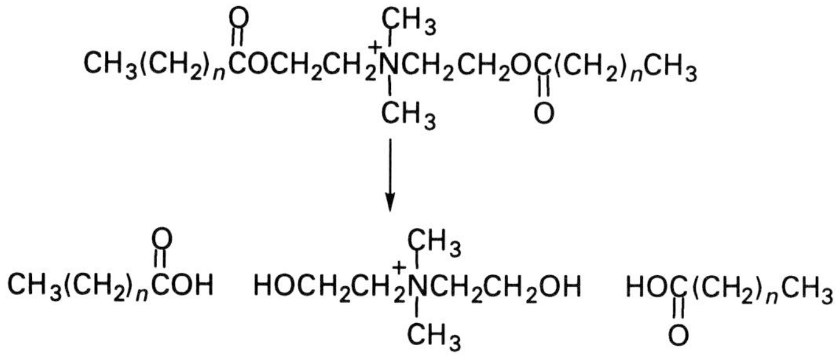

Figure 3.2 Biodegradation of a surfactant.

Hydrophobe structural features. The twin hydrophobes or 'tails' of a typical softener have four primary elements, which can be varied to meet the requirements of the application. The primary elements are the number of carbon atoms, the total degree of saturation, the quantity of polyunsaturates and the *cis* to *trans* ratio of the points of unsaturation [24–28, 30, 31]. Additional elements include substitution with non-carbon, hydrogen or oxygen substituents. To date these have been of little commercial consequence and they will not be discussed further here.

The carbon atom number for fabric softener structures usually follows the distribution of common oleochemical feedstocks. The number average is 17.5 for animal-based products such as tallow [32]. This carbon atom number also appears to be optimal for many structural variances. The physical manifestation of this is not understood but may be linked to the structural order on the surface of the fiber. Decreasing the carbon number decreases softening and lowers the deposition onto the fabric. Increasing the carbon number increases the difficulty of formulation due to the formation of gels at concentrations higher than ~5%. Carbon atom numbers closer to 16 are common for hair conditioners [33].

The total degree of unsaturation of any alkylene functionality is measured by its iodine value [34]. The iodine value measures the equivalents of iodine added across all of the unsaturation points in a molecule. Table 3.2 shows the iodine value and 16 and 18 carbon atom percentages of typical feedstocks used in cationic surfactant manufacture [35].

Table 3.2 Iodine values and C-16/C-18 content of feedstocks for ester-based cationics

Oil source	C-16 content	C-18 content	Iodine value
Palm	34–47	45–70	38–56
Tallow	23–46	53–77	50–60
Rapeseed (low erucic)	3–6	77–95	100–120
Linseed	11–16	76–95	170–200
Soybean	7–11	73–95	125–140

The iodine value can be manipulated during manufacture by hydrogenating either the feedstock or its derivatives [24, 25, 36]. The total degree of unsaturation of cationic surfactants for softening of textiles has increased substantially due to the introduction of concentrated softeners in both the US and Europe. Prior to this change the typical softener was made from a fully saturated ditallow dimethyl ammonium chloride formulated at 3 to 7% in an aqueous media [5]. Efforts to minimize solid waste and to improve economics led to the development of concentrated softener products. The new concentrated versions of fabric softeners are formulated at 17 to 26% active cationic surfactant [19]. The dihydrogenated tallow dimethyl ammonium chloride product cannot be formulated in excess of 16% without the formation of a high viscosity gel either in the final product or during manufacture of the dispersion unless additives are used to inhibit formation of the lamellar phase (Stig Freiberg, private communication). Increasing the degree of unsaturation of the tallow backbone increases the concentration at which these materials can be formulated. The higher concentrations achievable with unsaturated materials are a consequence of increased liquidity of the higher iodine value material and the lowered tendency to form lamellar phases due to the steric resistance against compacting unsaturated chains [24, 25].

The level of polyunsaturates in the hydrophobe of a cationic surfactant influences its liquidity and also its resistance to oxidative degradation and color formation [24, 37]. The higher the polyunsaturate level, and consequently the iodine value, the higher the liquidity and the higher the aqueous concentration of the softener dispersion that can be achieved. However, levels of polyunsaturates reduce the stability of the surfactants, particularly as related to color and odor [24, 25]. The development of color on storage in air is a well-known effect observed in the fats and oils that are precursors to cationic surfactants. The color is due to the propensity of polyunsaturated materials to react with oxygen and further propagate free radical reactions versus monounsaturated or fully saturated materials [37]. Products with high or even modest degrees of unsaturation frequently require the addition of an antioxidant such as the hindered phenol derivatives, butylated hydroxytoluene, BHT, and butylated hydroxyanisole, BHA [24, 25, 38].

The final element of the hydrophobe that can be manipulated is the *cis/trans* ratio of the unsaturated hydrocarbon fragments. As exemplified by the liquid nature of *cis*-stilbene, melting point –5°C, and *trans*-stilbene, melting point +124°C, [39] the higher the *cis* content the more liquid the product will be, all other things being equal. Natural tallow has a *cis/trans* ratio of about 8 to 20 [40]. Metal-catalyzed hydrogenation of fats and oils results in a reduction of the *cis/trans* ratio and an increase in the melting point of the oil when compared to a material of similar iodine value and a higher *cis/trans* ratio [41]. For concentrated fabric softeners high *cis/trans* ratios are preferred to reduce the likelihood of gel formation in the final product or during processing [24, 42–44].

The manipulation of each of these elements will influence the final product properties and can be used to customize the molecule to the requirements of the softener system. Typically, a combination of these elements can best satisfy the needs of the softener composition. This combination can be best optimized by the use of experimental design techniques where each of the elements can be varied independently but the influence of each upon the other can be evaluated [45].

Hydrophile structural elements. The hydrophile structure is the structural feature of the softener molecule that is most often manipulated in the development of new fabric softeners. The classic hydrophile for the traditional cationic softener is a positively charged nitrogen directly connected to the hydrophobe, as shown in Figure 3.3 [46, 47].

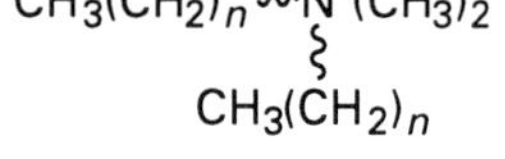

Figure 3.3 Classic hydrophile for the traditional cationic softener.

The new, more biodegradable, cleavable structures have the hydrophobe connected to the positively charged nitrogen hydrophile via an ester linkage, as shown in Figure 3.4. The structure of the hydrophobe in these materials can be more varied than with the traditional quaternaries.

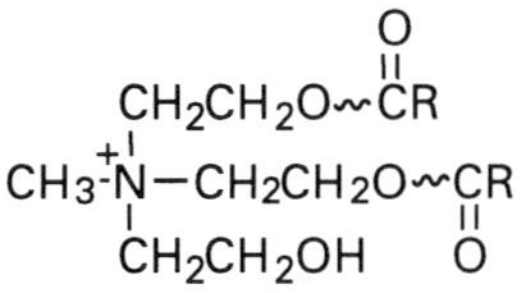

Figure 3.4 Structure of a biodegradable cationic softener.

Table 3.3 shows the common hydrophiles employed in ester-based cationic softeners. These structures are derived, for example, from an ethanolamine or from the reaction product of a chloro-substituted acid with a methyl amine. The hydrophile influences the behavior of the softener primarily by its size, which can alter the packing in the vesicular structure of a formulated material, and also by influencing the hydrophilicity of the fabric it is deposited upon.

Table 3.3 Structures of esterquaternary hydrophiles

Hydrophile structure	Reference
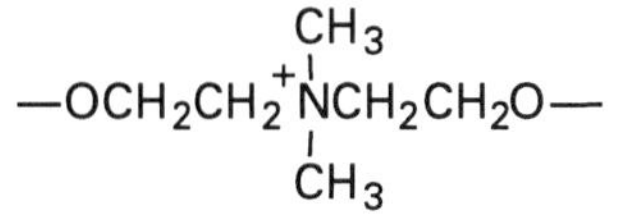	[24]
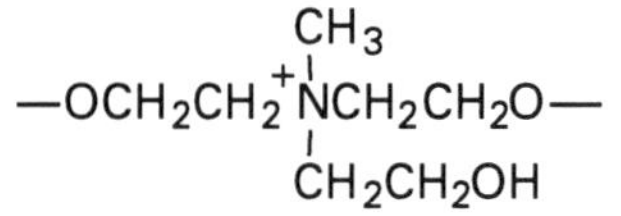	[18, 48, 49]
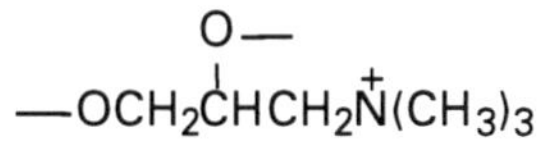	[50]
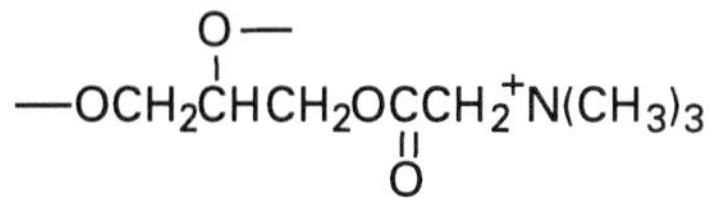	[29]
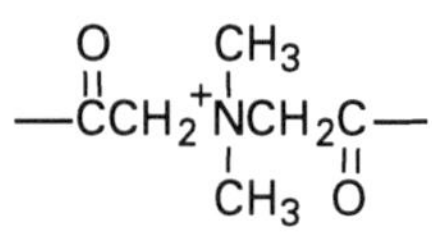	[23]
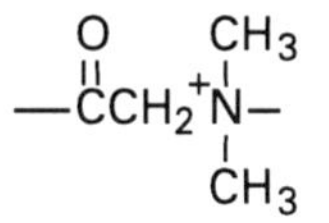	[51]

3.3.1.2 Formulations. In the typical fabric-softening formula the surfactant exists in a vesicular structure and the number of bilayers is a function of the concentration of the cationic surfactant in the solution [52]. The quality of dispersion influences the softening observed on the fabric. A poor dispersion leads to an uneven coating of softener on the fabric and insufficient deposition [50]. Cationic surfactants, which do not readily form stable vesicular structures, can be made stable by the inclusion of appropriate stabilizers, including ethoxylated alcohols [53] although other materials such as esterified glycerol esters have also been suggested for this use [54]. Recently, microemulsion formulae of fabric softeners have appeared [55]. These are made from water, glycol, dioleyl diester quaternary and an appropriate solvent. Some microemulsion formulations are given in Table 3.4. In addition, clear fabric softener formulae have been prepared from ester quaternaries, a C-8–C-22 mono-alkyl cationic surfactant and 17–75% of a non-aqueous organic solvent [56].

Table 3.4 Microemulsion softener formulae

Material	Formula 1 (wt%)	Formula 2 (wt%)	Formula 3 (wt%)
Water	57.5	57.5	57.5
Hexylene glycol	20		
Ethylene glycol monobutyl ether		20	
Butanol			20
Dioleyl diester quaternary	22.5	22.5	22.5
Appearance of solution	Clear	Clear	Clear
Appearance on dilution	Turbid	Clear	Turbid
Stability	Stable	Stable	Stable

3.3.2 Dryer softeners

Dryer softeners are employed primarily in North America to control the build-up of static charge in clothes tumble dryers. The product consists of a 17.5 × 30 cm polyester sheet. A mixture of a cationic surfactant and a 'release' or 'distribution' agent is applied to the sheet [57–59]. The level of the agent is typically 1.0 to 2.0 g per sheet and is in a solid form on the sheet [60, 61]. The sheet is tossed into the dryer with the wet clothes at the beginning of the cycle and controls the development of static charge throughout the dryer cycle.

The cationic surfactants used in dryer sheets are similar to those used in liquid softener compositions except that the counterion must be methosulfate. The methosulfate counterion is necessary due to the corrosivity of chloride ion to the internal parts of the dryer. The lack of concern about formation of lamellar gels and flowability in dryer sheet formulae leads to cationic surfactants for this application typically having hardened hydrophobes, unlike their liquid softener counterparts [57]. Both traditional nitrile-based cationic surfactants and ester-based cationics are suggested for use in this application.

From Figure 3.5, it can be seen that the removal of quaternary from the tumble dryer sheet is a result of the melting of the active component from the sheet, as well as a consequence of mechanical action or dissolution by the water contained in the wet fabric [18, 60]. As a consequence the melt characteristics of the softener active component are important in determining its function. The melt characteristic is determined from the cationic surfactant employed and also from the selection of the release or distribution agent.

Figure 3.6 shows the effect of altering the release agent on the transfer of the cationic surfactant to the clothes under use conditions. The higher mole ethoxylate causes a more rapid release as a result of the lowering of the melting point of the blend on the tumble dryer sheet. The specific control of the release can be fine-tuned by the use of experimental design techniques such as factorial analysis.

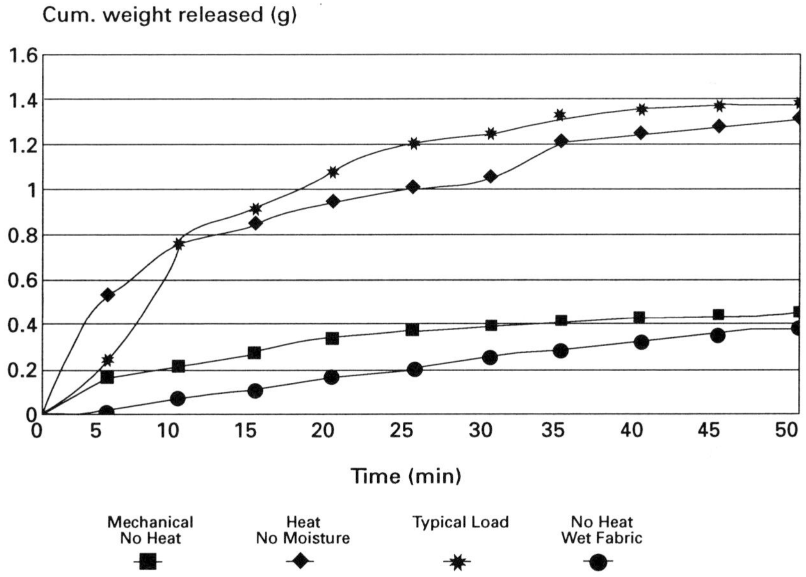

Figure 3.5 Determination of the removal of cationic surfactant from a tumble dryer sheet as a function of heat, dissolution and mechanical action.

Additional components of the tumble dryer sheet formula can include perfumes for marketing and soil-release agents to prevent soiling of fabric during the use period prior to the next washing [45].

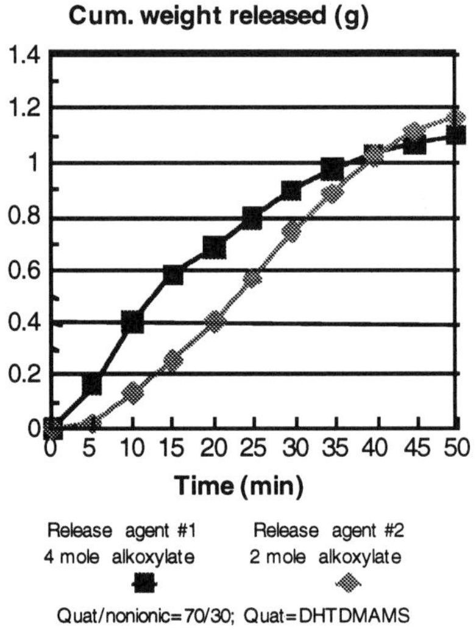

Figure 3.6 Measurement of the effect of the degree of alkoxylation on the release of cationic surfactant from a tumble dryer sheet.

3.3.3 Softergents

Cationic surfactants are sometimes included in detergent formulas as softening agents. These softergents provide ease of use but typically do not provide the cleaning level of detergent formulae without these materials. One example of a heavy duty laundry detergent contains lauryl trimethyl ammonium chloride as the softening agent in which C-12 alkyl benzene sulfonate is specifically excluded or limited to prevent phase separation and formation of a translucent viscous formula [62, 63]. Inclusion of a C-8–C-20 fatty acid into these formulae provides enhanced softening and cleaning.

3.3.4 Hair conditioners

The washing of hair with synthetic detergents has a similar effect to the washing of clothes. In the case of hair the natural oils that normally coat hair are washed away and leave the hair with a lack of shine, tangles and a condition known as 'flyaway'. Flyaway, as the name implies, is caused by hair fibers repelling from one another due to the build-up of static charge. This results in the hair having an unkempt appearance. Conditioners restore the hair to its original condition by providing a synthetic oil to repair the damaged hair fibers. Cream rinse hair conditioners contain approximately 2% of a hydrophobic cationic surfactant. The hydrophobic cationic surfactants that are used in hair conditioners are shown in Table 3.5.

Table 3.5 Cationic surfactants used in hair conditioning products.

Cationic surfactant	Reference
Dodecyl-trimethyl ammonium chloride	[64–66]
Di-(palmitoyloxyethyl)-2-hydroxyethyl-methyl ammonium chloride	[32]
Di-(hydrogenated-soyoyloxyethyl)-2-hydroxyethyl-methyl ammonium chloride	[18]
Di-cetyl dimethyl ammonium chloride	[67]
Di-(hydrogenated tallow)-dimethyl ammonium chloride	[68]
Tri-cetyl-methyl ammonium chloride	[69]

The cationic surfactant is selected on the basis of the amount of conditioning desired, the more hydrophobic cationics, such as di-(hydrogenated tallow)-dimethyl ammonium chloride, provide a high level of conditioning, while the same molecular architecture based on dipalmityl-dimethyl ammonium chloride provides a lower level of conditioning. Still milder conditioning for oily hair can be provided by mono alkyl chain cationics such as dodecyl-trimethyl ammonium chloride.

Developments in this application of cationic surfactants follow the developments in the fabric softener field since traditionally fabric softener surfactants have been employed in hair conditioners. The use of ester-based cationics is drawing attention. Again, the selection of the hydrophobe controls the performance of the cationic surfactant. Figure 3.7 shows the combing force performance of an ester-based cationic, di-(hydrogenated-soyoyloxyethyl)-2-hydroxyethyl-methyl ammonium chloride (EQ), compared to the traditional quaternaries, di-(hydrogenated tallow)-dimethyl ammonium chloride (Di-HT) and tri-cetyl-methyl ammonium chloride (Mono 16), determined using a standard Diastron testing protocol [70]. The conditioning performance of the ester-based cationic is excellent and improved static control is also demonstrated. This is probably due to the improved hydration of the hair follicle as a consequence of the more polar nature of the ester-based cationics.

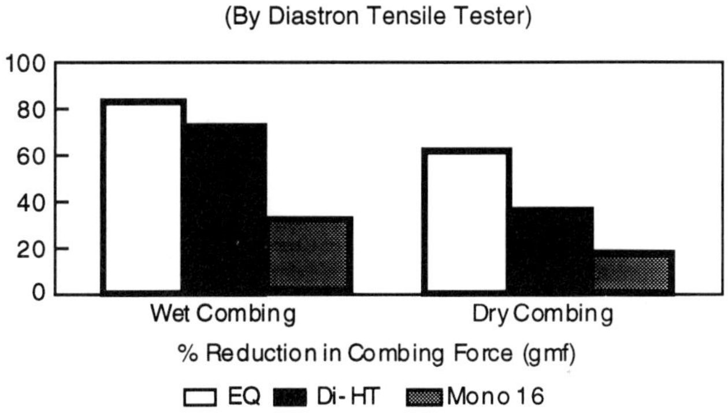

Figure 3.7 Comparison of human hair combing force for cationic surfactants.

3.3.5 Detergents

Cationic surfactants are not used as the primary surfactant in laundry detergents as this responsibility falls to the commodity surfactants such as linear alkyl benzene sulfonates, ethoxylated alcohols and alkyl sulfates. Nevertheless, over the past 20 years there has been extensive work claiming that the use of cationics improves the performance of their anionic counterparts [71–73]. A combination of an alkyl phosphate (APE) and tetradecyl-trimethyl ammonium bromide (TTAB) substantially lowers the interfacial surface tension between the surfactant solution and a variety of oils compared with either surfactant itself, as shown in Table 3.6 [72].

Table 3.6 Interfacial tension of anionic/cationic surfactant mixtures (dynes/cm)

Surfactant	APE	APE/TTAB	TTAB 1 g/l
Oil	1 g/l	4:1 0.3 g/l total	1 g/l
Hexadecane	3.0±0.3	1.1±0.2	4.9±0.2
Nujol	2.6±0.3	1.1±0.2	2.3±0.1
Dirty motor oil	1.3±0.7	0.8±0.1	1.4±0.2
Wesson oil	3.9±0.6	1.4±0.2	2.8±0.6
Oleic acid	4.9±0.9	4.7±0.6	4.6±0.6

TTAB in combination with an alkyl ethoxylated sulfate also lowers the surface tension observed at the critical micelle concentration (CMC) by approximately 10 dynes/cm compared with either of the surfactants alone. It is postulated that the combination of the anionic and cationic results in charge neutralization, providing a 'net' surfactant that is similar to a non-ionic surfactant. This is confirmed by the lowering of the cloud point of the surfactant combination as the mole fractions of the anionic and cationic components are adjusted to near the equivalency point.

3.3.6 Thickeners

As a consequence of their tendency to form lamellar phases or rod-shaped micelles at low concentrations, cationic surfactants are frequently employed as the primary surfactants to thicken high-salt formulae [74, 75]. The viscoelastic nature of certain cationic surfactant solutions has been employed in a novel way to allow a solution of sodium hypochlorite not to be easily diluted and therefore to remain at a higher concentration for the purpose of oxidizing the clogs of human hair that form in drains [75]. Low concentrations of tri-cetyl-methyl ammonium chloride in combination with two hydrotropes form viscoelastic solutions with the values of viscosity and shear stress/modulus, Tau/G_0 shown in Table 3.7.

Table 3.7 Properties of tri-cetyl-methyl ammonium chloride solutions

Tri-cetyl-methyl ammonium chloride (wt%)	Sodium xylene sulfonate (wt%)	Chlorobenzoic acid (wt%)	Viscosity (cP)	Tau/G_0 (s/Pa)
0.370	0.260	0.080	47	0.35
0.5000	0.143	0.071	247	0.45
0.625	0.125	0.063	716	0.89
0.625	0.250	0.063	140	0.06

3.3.7 *Hard-surface cleaners*

Cationic surfactants, because of their positive charge, are attracted to many surfaces, altering the surface and changing the force required for cleaning. Table 3.8 shows the effect of two cationic surfactant formulae when pre-applied onto a tile prior to soiling [76].

Table 3.8 Effect of cationic surfactants on effort required for cleaning

	Composition number			
	1	2	3	4
Non-ionic (ethoxylated alcohol)	10%	10%	10%	10%
2-amino-2-methyl-1-propanol	4%	4%	4%	4%
K_2CO_3	1.2%	1.2%	1.2%	1.2%
Digol	8%	8%	8%	8%
Propylene glycol			0.6%	3.0%
Tri-cetyl-methyl ammonium bromide		1%		
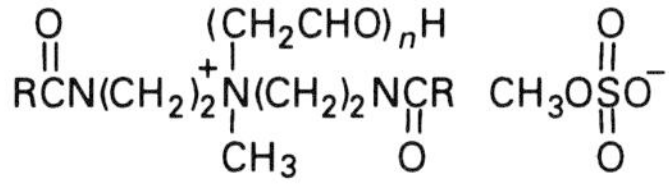 R= saturated alkyl			0.2%	1.0%
Effort Ns (Newton seconds)	930	469	189	137

It is believed that the cationic surfactants modify the surface energy, which raises the contact angle of the soil subsequently deposited on the surface. This allows for more effective cleaning with less effort as opposed to an untreated tile or one treated with a composition without a cationic surfactant.

3.3.8 *Cationic surfactants in polymerization*

Although polymerization in microemulsions has been known for some time [77], only recently has information on the effect of surfactant chain length on the cationic-stabilized systems become known. Studies of the polymerization of methacrylate derivatives in bicontinuous microemulsions stabilized by *n*-alkyl trimethyl ammonium bromide reveal that it is possible to control polymer porosity and microstructure by the proper choice of surfactant chain length [78]. In an extension of this work it was shown that combinations of long- and short-chain *n*-alkyl trimethyl ammonium bromides influence the morphology and microstructure of the polymer [79]. Increasing the long-to-short-chain ratio results in an increase in pore size.

The three-component butyl methacrylate/water/dodecyl trimethylammonium bromide microemulsion [80] is reported to give very fast reaction rates and high conversions regardless of the initiator type. The microemulsion system vinyl-toluene/cetyltrimethyl ammonium bromide/water has also been investigated [81]. In each of these systems similar particle sizes (~30 nm) and molecular weights (~3×10^6 Da) are observed. Interestingly, only a few polymer chains are required per particle.

Another use of cationic surfactants in polymer systems involves polymerization of the cationic surfactants. Studies of this chemistry may be separated into two classes: head polymerization and tail polymerization. Head polymerization involves a moiety attached near the polar end of the amphiphile whereas tail polymerization requires participation of the hydrophobe. Quaternization of dimethylaminoethyl methacrylate with alkyl halides gives head-polymerizable cationic surfactants. The surface properties of these products have been studied as part of an investigation of nanoencapsulation technology [82]. The behavior of quaternary derivatives of dialkylamino methacrylates with differing alkyl groups was studied to determine the effect of changing the chain-length ratios. As might be expected, changing the ratios of the alkyl chains affects the aggregation of the surfactants and thereby also affects rates of polymerization [83]. Head-polymerizable allylic-containing quaternary salts have been patented for use as emulsifiers and dispersants in suspension polymerizations [84].

Tail polymerization of cationic surfactant monomers is illustrated by the copolymerization of 11-(acryloyloxy)undecyl trimethyl ammonium bromide with styrene in an aqueous microemulsion to give a nanolatex with particles of approximately 21 nm [85]. These particles are redispersed readily, indicating that the polar groups are in the outer shell of the particles. The same polymerizable surfactant has been used to prepare ultrafiltration membranes by microemulsion copolymerization with methyl methacrylate and 2-hydroxyethyl methacrylate [86]. Studies of a similar surfactant, 11-(methacryloyloxy)undecyl trimethyl ammonium bromide, were run in aqueous and organic media. Non-polar organic solvents favor higher chain growth [87].

The α-olefin moiety of 10-undecenyl trimethyl ammonium bromide has been exploited to prepare a polymeric surfactant that is more water soluble than the surfactant monomer. This work compares the surfactant properties of 10-undecenyl trimethyl ammonium bromide and dodecyl trimethyl ammonium bromide [88].

3.3.9 Organoclays

Production of organoclays is a large market for quaternary ammonium salts [16]. These products, prepared by ion exchange between the clay platelets and the quaternary ammonium salt, are useful for modifying and controlling the rheology of non-aqueous fluids such as lubricating oils, linseed oil and toluene [89]. The use of organoclays to control rheology of non-aqueous systems provides greases, solvent-based paints and drilling fluids. For these applications, dimethyl di(hydrogenated tallow alkyl) ammonium chlorides or dimethyl benzyl (hydrogenated tallow alkyl) ammonium chlorides are the quaternary salts most commonly used, in combination with smectite-type clays. For more polar systems, i.e. containing polar organic solvents or water, poly-ethoxylated quaternary ammonium salts are needed. These products have been used to thicken paints [90] and to gel nail polish [91]. An interesting recent development is the preparation of a novel nanocomposite of clay and polymer by exchanging cationic surfactant monomer onto the clay platelets then polymerizing these organoclay platelets with acrylamide [92].

An application of organoclays with great potential is the removal of contaminants from water by adsorption. This is the subject of studies of the adsorption of phenolics onto hexadecyl trimethyl ammonium-modified montmorillonite [93] and adsorption of organics from coal-gas washings onto hexadecyl trimethyl ammonium-modified bentonite [94]. Using Raman spectroscopy to study the interaction of hydrocarbons with cetyl trimethyl ammonium bromide (CTAB)-modified montmorillonite it was shown that the quaternary salt was in a liquid-like state. [95]. It was concluded that interaction of organic compounds in this system is best classified as absorption.

A potential problem during effluent remediation by adsorption organic contaminants onto organoclays is release of the quaternary salt. This loss of quaternary salt has been shown to be minimal [96].

3.3.10 Miscellaneous

3.3.10.1 Drag reduction. In the presence of sodium salicylate, tris(hydroxyethyl) tallow alkyl ammonium acetate is an efficient drag-reduction agent [97]. The transient behaviors for the normal stress difference and viscosity of this system are dependent on the ratio of sodium salicylate to tris(hydroxyethyl) tallow alkyl ammonium acetate. At initial start-up, a maximum for the normal stress difference and viscosity is attained in the quaternary:salicylate ratio range of 2.5–3; however, at steady state, not much variation is observed over the ratio range of 1.5–3.

3.3.10.2 Soil/groundwater treatment. Adsorption of hexadecyl trimethyl ammonium surfactant onto sand was used to model the sorption of chlorinated hydrocarbons from water into a surfactant-treated medium [98]. However, a similar study using the same cationic surfactant revealed that low surfactant concentrations persist in the treated zone even after extensive flushing [99]. This may have consequences for microorganisms within the aquifer. In general, these systems are intended to prevent spreading of contaminants thereby allowing for localized redemption.

3.3.11 Surfactant recovery

Surfactant precipitation by addition of oppositely charged ionic additives is claimed to be an effective recovery method [100] but this requires addition of another reagent whose disposal must be taken into account. A better method, which might allow for recycling of the cationic surfactants, is foam fractionation. This also might allow reduction of the surfactant concentration to environmentally acceptable levels [101].

3.4 Developments in cationic surfactants

3.4.1 Cationic surfactant/non-ionic polymer systems

Recent publications in this area cover a diverse range of polymer substrates. EMF, electromotive force, [102] and calorimetric [103] methods have been used to investigate the interaction of CTAB with non-ionic polymers. EMF has been used to examine polyethylene glycol and polypropylene glycol as substrates, whereas the calorimetric approach also investigated, in addition to polypropylene glycol, poly(vinylmethylether) and ethyl(hydroxyethyl) cellulose. Both studies support the concept of attachment of surfactant aggregates to the polymer backbone. The interaction of CTAB and cetylpyridinium chloride with copolymers of N-isopropylacrylamide and N-*n*-octadecylacrylamide tends to occur by adsorption of individual surfactant molecules in a partitioning mechanism before micelle formation [104].

A comparison of the interactions of sodium dodecyl sulphate, SDS, and DTAB (dodecyl trimethyl ammonium bromide) with hydrophobically modified ethoxylated urethane associating polymers was carried out using nuclear magnetic resonance (NMR) techniques [105]. This study also contrasted the surfactants' interaction with poly(ethylene oxide).

Cationic and non-ionic surfactants interacting with hydrophobically modified polyols have been discussed as rheological agents for cleaners [106].

3.4.2 Cationic surfactant/cellulosic systems

The influence of cationic surfactants on the aqueous behavior of cellulosics has been reported for three classes of cellulose derivatives: non-ionic, anionic and cationic. For the non-ionic cellulose derivatives, ethyl(hydroxyethyl)cellulose (EHEC) and methylcellulose (MC), differences in behavior toward CTAB are found [107]. EHEC interacts strongly with the surfactant whereas MC is slightly interactive. The strength of interaction modifies the phase separation (clouding point) that occurs when aqueous solutions of these polymers are heated. Thermoreversible gels can be formed from EHEC/cationic surfactant combinations [108].

Anionic sodium (carboxymethyl)cellulose (NaCMC) interactions with alkyl trimethyl ammonium bromide were investigated using a surfactant-sensitive electrode and fluorescence quenching [109]. The effects of added electrolyte in the binding to NaCMC were reported, and estimates of the critical aggregation concentrations and binding cooperativity given.

The chloride salt of the N,N-dimethyl-N-dodecyl derivative of hydroxyethylcellulose has been shown to interact with alkyl trimethyl ammonium halides through hydrophobic interactions that are sufficiently strong to overcome unfavorable electrostatic repulsions [110]. These interactions are similar to hydrophobic interactions of hydrophobically modified anionic and non-ionic water-soluble polymers. Another study of surfactant interactions with cationic cellulosics is available, although the primary focus of the study is anionic surfactants [111].

3.4.3 Cationic surfactant/charged polymer systems

A general study of the interaction of polyelectrolytes with oppositely charged surfactants has been published [112]. Most work in this area centers on interactions of anionic polymers with cationic surfactants. Acrylic derivatives are most commonly reported [113–117]. Other systems studied include polyacrylamide sulfonates [118, 119], partially hydrolyzed polyacrylamide [120], maleic acid copolymers [121, 122] and poly(sodium styrenesulfonate) [123]. Potassium styrenesulfonate copolymers combined with cationic surfactants are patented for use in antistatic coatings for polyester films [124].

Cationic-surfactant-modified fluorosulfonic acid polymers have been claimed to be of use in electrostatic clutch systems [125]. Membranes for the separation of hydrocarbons from alcohols are prepared from sodium cellulose sulfate and alkyl pyridinium chlorides. Typical separations involved hydrocarbons and lower alcohols [126].

The stabilizing interaction of cationic polyelectrolytes, quaternized poly-(2-vinylpyridines), with octadecyl dimethyl ammonium chloride in concentrated calcium chloride solutions has also been reported [127].

3.4.4 Inorganic cationic surfactant systems

The equilibrium of CTAB on mica has been studied using the Whihelmy plate method [128]. In this study solid/liquid and solid/vapor isotherms are given along with a description of the solid/liquid bilayer. The frictional properties of double-chain quaternary ammonium salt monolayers on mica were analyzed using atomic force microscopy (AFM). By changing the AFM probe tip, the authors were able to demonstrate a relationship between nanoscale and microscale frictional properties [129]; this has implications for tribology and lubrication.

Another investigation showed that the long-range force created by interacting dodecylamine hydrochloride with mica surfaces is pH dependent [130]. This study also discusses the implications of this phenomenon for mineral flotation.

Solubilization of toluene into cationically modified vermiculite has been accomplished using cationic surfactants of different chain lengths [131]. Differences in swelling of the intercalated systems were analyzed in terms of hydrophobicity and structural disorder.

A review of the adsorption of cationic surfactants on various mineral oxides has been published [132]. The review discusses adsolubilization of solutes and some industrial applications. A significant number of related publications dealing with interactions of quaternary ammonium salts with silica, alumina or titanium dioxide has appeared [133–143]. An interesting study of the interaction of aluminate and silicate with cetyl trimethyl ammonium chloride (CTAC) has revealed that the aluminate has little effect on CTAC aggregation [144]. On the other hand, silicate ion, from sodium silicate or tetramethyl ammonium silicate, has a strong influence on surfactant aggregation. Initially flexible, worm-like micelles are enclosed in silicate oligomer, then further condensation of silicate leads to hexagonal liquid crystals. This phenomenon is important in formation of the MCM-41 family of mesoporous molecular sieves. A solid-state ^{13}C NMR characterization of CTAC/mesophase silicate systems is available [145].

3.4.5 Traditional cationic surfactants

Traditional cationic surfactants are those containing N-fatty alkyl groups. Although these may be mono-, di-, tri- or even tetra-alkyl systems, the majority of commercial products are the mono- and dialkyl species. In addition, most commercial cationic surfactants, being derived from natural sources, have mixed alkyl chain lengths, yet most studies of surfactant properties use single chain length substrates. This sometimes leads to uncertainty in predicting the behavior of commercial products when using published data. The situation has been clarified significantly by a recent study of pure, binary and ternary mixtures of the series, i.e. dodecyl, tetradecyl and hexadecyl trimethyl ammonium bromides and hexadecyl pyridinium chloride [146]. In addition to providing data on CMC (critical micelle concentration), counterion binding, aggregation numbers, etc., the work revealed that mixed systems are non-ideal but have similar micelle polarities and energy parameters compared to the single chain length surfactants and binding of counterions is less for the mixed systems than for pure states. Another study of binary mixtures found that addition of dodecyl trimethyl ammonium bromide to aqueous hexadecyl trimethyl ammonium bromide lowers the CMC for hexadecyl trimethyl ammonium bromide, as the mole fraction of dodecyl trimethyl ammonium bromide is increased, and also stabilizes the micelles entropically [147]. Data on the effect of temperature on the aqueous properties of hexadecyl trimethyl ammonium bromide are given for four temperatures (18, 25, 35 and 45°C). Other information of value in this research includes densities, viscosities and surface tensions. Additional data on the interaction of dialkyl salts with monoalkyl salts is contained in a study of the easy penetration of aqueous dioctadecyl dimethyl ammonium bromide by monoalkyl trimethyl ammonium bromides, especially when the alkyl chain lengths of the quaternary salts are similar [148].

An extensive report of the CMCs, surface tensions and head group areas for dialkyl dimethyl ammonium bromides with differing alkyl groups shows that, as expected, CMC values decrease with increasing chain length [149]. Interestingly, at surfactant concentrations above the CMC, surface tension decreases and gives a second break. This is taken to indicate a change from globular to rod-like micelles. Surfactant headgroup areas and interfacial tension measurements against decane are reported. Additional information on mixed alkyl systems is found in a study of water-in-oil microemulsions containing 'alkyl' dodecyl dimethyl ammonium bromides where 'alkyl' is C-10, C-12, C-16 or C-18 [150].

The behavior of pure dialkyl quaternary salts continues to be of interest, as evidenced by recent studies reporting the concentration dependence of vesicle formation by didodecyl dimethyl ammonium bromide [151] and the formation of dioctadecyl dimethyl ammonium chloride/water/hydrocarbon/fatty alcohol microemulsions [152]. The fatty alcohols used were 1-decanol and 1-dodecanol; the hydrocarbons included octane, decane, dodecane, tetradecane and hexadecane.

3.4.6 Influence of anions

It is well known that electrolytes have an influence on the behavior of cationic surfactants. The practical implications of this phenomenon have been published in a broad study of the effects of synthetic hard river water on surfactant properties for decyl, dodecyl and tetradecyl trimethyl ammonium bromide, as well as decylpyridinium bromide [153]. As expected the surface activity increases in the presence of electrolytes. The study also gives a significant amount of data comparing similar cationic, non-ionic and anionic surfactants. While this work provides information on behavior in hard river water, two other studies address the environmental fate and impact of cationic surfactants. One study provides a review of the biodegradability of cationic surfactants [154]. The other study reveals the effect of the length of the alkyl chain on the aquatic toxicity of alkyl trimethyl ammonium and alkyl benzyl dimethyl ammonium halides [155].

3.4.7 Symmetrical diquaternary surfactants: gemini surfactants

This interesting class of surfactants is essentially unexploited commercially. An easily accessible type of gemini surfactants is that derived from epichlorohydrin. It has been shown that during reaction of epichlorohydrin with N,N-dimethyldodecylamine and its hydrochloride, which assists ring opening of the epoxide ring, neighboring group participation by the hydroxyl group is important in the quaternization [156]. A continuation of this work reveals that these products have typical surfactant properties and are potentially good foaming agents [157]. As expected the CMCs of these diquaternary salts decrease with increasing chain length of the alkyl group. However, the CMCs are two orders of magnitude below those for the corresponding monoalkyl monoquaternary salts. Interestingly, with the exception of foaming properties, the products within this class of diquaternary salts do not seem to be sensitive to the connecting group between the quaternary centers. In contrast, for gemini surfactants based on dodecyldimethylamine, the growth and shape of the micelles are dependent on the spacing between the quaternary centers [158]. The increase of the spacing from ethylene to hexylene progressively hinders the growth of micelles. Shorter spacing groups give more variation in the growth and shape of the micelles. This dependence of surfactant properties on the length of the spacing between the quaternary centers has been demonstrated using small-angle neutron scattering and fluorescence probe analysis [159].

An investigation of the surfactant properties of a series of gemini surfactants with 2,3-dihydroxybutenyl linking groups in water and in dilute sodium chloride and sodium bromide brines shows that, for certain alkyl chain lengths and depending on the aqueous environment, the surfactant activities are less than expected [160]. Formation of aggregates below the CMC are proposed. Confirmation of the lower CMC values for gemini surfactants compared to the corresponding monoquaternary salts has been carried out [161] but it was found that surface tensions of the gemini systems are not very different from the mono systems. Interestingly, gemini surfactants have better oil solubilization, especially for aromatic compounds, than mono cationics. The viscoelastic properties of gemini surfactants also differ from those of monoquaternary salts.

A comparison of the adsorption of monoquaternary, diquaternary (gemini) and triquaternary salts containing one, two and three dodecyl chains, respectively, on to silica revealed that the amount of surfactant adsorbed at saturation decreases in the sequence mono, gemini, triquaternary [162]. The order of the ratio of maximum adsolubilized 2-naphthol to adsorbed surfactant follows the opposite trend. The aqueous aggregation properties (CMCs, aggregation numbers) of this set of surfactants are given elsewhere [163].

An unusual class of gemini surfactants derived from arginine is reported to be bioactive [164]. A brief review of gemini surfactants is available [165].

3.4.8 Catanionic surfactants

This interesting class of surfactants, which will be increasingly investigated in the future, is derived from the combination of amphiphilic anions and cations. Although not specifically cationic in nature, they do represent a novel, as yet unexploited, application of cationic surfactants. As synthesis of catanionic surfactants is usually done by simply mixing the cationic surfactant with the anionic surfactant, detailed analysis of the behavior of these surfactants must take into account the formation of an equivalent amount of salt, e.g. sodium bromide.

Microemulsions and other phase behaviors of SDS and CTAB have been reported for a system containing heptane, alcohol and water [166]. The effects of alcohol chain length and surfactant ratios are given. Combinations of anionic and cationic surfactants show different solubilization capacities toward hexanol compared to heptane [167]. When mixed in ratios other than 1:1 (cationic:anionic), catanionic systems can exhibit phase separations, with one phase being near the optimum 1:1 combination. Dodecyl triethyl ammonium bromide and SDS have been use to demonstrate this phenomenon [168]. Fluorescence probes have been used to show that the separate phases have significantly different compositions and properties. A fluorescence probe was also used to study alkyl carboxylate/alkyl trimethyl ammonium systems [169].

It has been shown, using the sodium octylsulfate and CTAB catanionic combination, that electrostatic interactions are important in determining phase behavior [170]. Addition of electrolyte (NaBr) strongly affects aggregation behavior and the vesicle–micelle transition. Indeed, calcium chloride affects the precipitation of anionic surfactants and catanionic mixtures [171].

A thermodynamic treatment of vesicle formation in the catanionic regime has been presented for the system SDS/dodecylammonium chloride [172]. Calculations indicate that the work of bending a bilayer plane to a closed bilayer vesicle increases as the ratio of anionic surfactant to cationic surfactant approaches unity. Vesicle formation is most difficult at the 1:1 ratio. The unique behavior of catanionic systems has also been demonstrated in a study of various types of binary surfactant combinations [173]. The effect of the symmetry of the cationic surfactant, i.e. the ratio of the hydrophobic tails, is reported for dodecyl trimethyl ammonium chloride in systems containing either perfluorinated or hydrocarbon tails in the anion [174]. Further insight into the phase behavior of catanionic systems is provided in a study of the bis(2-ethylhexyl)sulfosuccinate/didodecyl dimethyl ammonium bromide/water system [175]. Although precipitation of the catanionic surfactant occurs at concentrations between 1 and 10 wt% in water, concentrations above this value lead to redissolution.

Cation and anionic surfactants have been shown to penetrate the bilayers of vesicles of dimethyl dioctadecyl ammonium bromide and of sodium didodecylphosphate [176]. Gemini catanionic surfactants derived from diquaternary salts and alkylsulfates or alkylethersulfonates [177] have been shown to have unusual surfactant properties, depending on the alkyl chain lengths in the hydrophobes. Synergies on the reduction of surface tension have also been discussed [177].

The adsorption of a gemini surfactant, 1,2-bis(dodecyldimethylammonio)-ethane dibromide, on silica in the presence of sodium bromide has been shown to be increased by increasing the amount of added sodium bromide [178]. This study also discusses the effect of added sodium bromide on the ability of the gemini surfactants/silica system to adsolubilize 2-naphthol.

3.4.9 Cationic/non-ionic surfactant systems

Studies of the adsorption and monolayer structure of the binary system hexadecyl trimethyl ammonium bromide/monododecyl hexaethylene glycol indicate that temperature-induced dehydration of the surfactants plays an important role in determining the nature of monolayer at the air–water interface [179, 180]. Another mixed system, tetradecyl trimethyl ammonium chloride/pentadecyl ethoxylated(15) nonylphenol, is proposed to give two different mixed micelles above certain surfactant ratio-dependent concentration ranges [181]. In an investigation of the co-adsorption of a somewhat similar binary system on to polytetrafluoroethylene, maximum adsorption was observed for the combination hexadecyl trimethyl ammonium bromide/nonylphenol ethoxylate(13) [182].

3.4.10 Calculated surfactant properties

A continual goal of surfactant scientists is a computational method for predicting properties and phase diagrams for surfactant systems. A further step in this direction is a computer program for the calculation of the aqueous properties of surfactants based on the molecular structures of surfactants [183]. The value of this method awaits further development. Properties available from this system include CMC, micelle shape and size, concentration required for phase separation, and surface tension. An empirical system using zeta potential measurements to determine Davies incremental hydrophilic-lipophilic balance, HLB, values for cationic surfactants provides an additional tool for more efficient surfactant selection [184].

References

[1] Anon. (1997) *Chemical Marketing Reporter*, May 19, p. 18.

[2] Gadberry, J.F. (1990) in *Cationic Surfactants: Organic Chemistry* (ed. J. Richmond), Marcel Dekker, New York and Basel.

[3] Ralston, A.W. (1948) *Fatty Acids and Their Derivatives*, Wiley, New York.

[4] Puchta, R., Krings, P. and Sandkuhler, P. (1993) *Tenside Surf. Det.*, **30** 186-91.

[5] Cahn, Arno (ed.) (1994) *Proceedings of the 3rd World Conference on Detergents: Global Perspectives*, AOCS Press, Champaign, IL.

[6] Fruth, A., Strauss, J. and Stuhler, H. (1992) US Patent 5,175,370, assigned to Hoescht.

[7] Fruth, A., Strauss, J. and Stuhler, H. (1993) US Patent 5,231,238, assigned to Hoescht.

[8] Wyness, G. and Jensen, D. (1968) US Patent 3,379,764, assigned to Reilly Chemicals.
[9] Eriksson, J.B. and Dadekian, Z.A. (1978) US Patent 4,100,210, assigned to Lonza A.G.
[10] Dadekian, Z.A. and Wilbourn, D.S. (1969) US Patent 3,432,561, assigned to Baird Chemical.
[11] McKinnie, B.G and Harrod, W.B. (1994) US Patent 5,347,053, assigned to Albemarle.
[12] Davis, W.T. (1977) US Patent 4,024,189, assigned to Ethyl Corporation.
[13] Goe, G.L., Keay, J., Scriven, E., Prunier, M. and Quimby, S. (1995) US Patent 5,424,436 assigned to Reilly Industries.
[14] Stanley, K.D. and White, K.B. (1986) US Patent 4,569,800, assigned to Akzo Nobel, Inc.
[15] Stanley, K.D. and White, K.D. (1987) US Patent 4,675,118, assigned to Akzo Nobel, Inc.
[16] Dery, M. (1997) *Quaternary Ammonium Compounds*, in *Kirk Othmer Encyclopedia of Chemical Technology*, 4th edn, vol. 20, John Wiley and Sons, Inc., New York.
[17] Brown, D.M., Gatter, E. and Littau, C. (1995) US Patent 5,463,094 assigned to Hoescht Celanese.
[18] Technical Data, Akzo Nobel Chemicals, Dobbs Ferry, NY.
[19] McConnell, R.B. (1994) *Inform*, **5** (1) 76
[20] Berenbold, H. (1995) *Inform*, **5** (1) 82
[21] McAvoy, White, C.E., Moore, B.L. and Rappaport, R.A. (1994) *Environmental Toxicology and Chemistry*, **13** (2) 213-21.
[22] ECETOC Technical Report No. 53, *DHTDMAC: Aquatic and Terrestrial Hazard Assessment*, CAS No. 61789-80-8, February, 1993, Brussels, Belgium.
[23] Wueste, B. Weissen, H.J. and A.G. Fischer (1997) US Patent 5,596,125, assigned to Akzo Nobel Inc.
[24] Wahl, E., Bacon, D., Baker, E., Bodet, J.-F., Demeyere, J., Severns, J., Siklosi, M., Vogel, A. and Watson, J. (1995) US Patent 5,474,690, assigned to Procter and Gamble.
[25] Bacon, D.R. and Trinh, T. (1996) US Patent 5,500,138, assigned to Procter and Gamble.
[26] Courdavault Duprat, S. *et al.* (1993) European Patent Application 550,361, assigned to Stepan.
[27] Godefroy, L. (1993) European Patent Application 580,527, assigned to Stepan.
[28] Uphues, G., Ploog, U., Jeschke, R. and Waltenburger, P. (1994) US Patent 5,296,622, assigned to Henkel.
[29] Weissen, H.J. and Porta, N. (1996) US Patent 5,543,066, assigned to Akzo Nobel.
[30] Yamamura, M. *et al.* (1991) US Patent 5,023,003, assigned to Kao.
[31] Mermelstein, R., *et al.* (1995) W.O. Patent 95/31524, assigned to Procter and Gamble.
[32] Sonntag, N.O.V., Composition and Characteristics of Individual Fats and Oils, in *Bailey's Industrial Oil and Fat Products*, vol. 1, 4th edn (ed. Daniel Swern), John Wiley and Sons, p. 343.
[33] Bonastre, N. and Subirana, R.Pi (1996) European Patent Application 739,976 assigned to Henkel.
[34] Sonntag, N.O.V., Reactions of Fats and Acids, in *Bailey's Industrial Oil and Fat Products*, vol. 1, 4th edn (ed. Daniel Swern), John Wiley and Sons, p. 115
[35] *Fatty Acids and Glycerine* (1996) product brochure, Akzo Nobel Chemicals.
[36] Sonntag, N.O.V., Reactions of Fats and Acids, in *Bailey's Industrial Oil and Fat Products*, vol. 1, 4th edn (ed. Daniel Swern), John Wiley and Sons, p. 113.
[37] Sonntag, N.O.V., Reactions of Fats and Acids, in *Bailey's Industrial Oil and Fat Products*, vol. 1, 4th edn (ed. Daniel Swern), John Wiley and Sons, pp. 137-53.
[38] Iacobucci, P. *et al.*, US patent applied for, assigned to Akzo Nobel, Inc.
[39] Windholz, M. (ed.) (1976) *The Merck Index*, 9th edn, Merck and Co., Inc., p. 1139.
[40] Swern, D. (ed.) (1979) *Bailey's Industrial Oil and Fat Products*, vol. 1, 4th edn, John Wiley and Sons, New York, pp. 344-45.
[41] Patterson, H.B.W. (1994) *Hydrogenation of Fats and Oils: Theory and Practice*, AOCS Press, Champaign IL.
[42] Severns, J.C. *et al.* (1996) US Patent 5,531,910, assigned to Procter and Gamble.
[43] Wahl, E.H. *et al.* (1994) W.O. Patent 94/20597, assigned to Procter and Gamble.

[44] Uphues, G. *et al.* (1994) US Patent 5,296,622, assigned to Henkel.
[45] Davies, O.L. and Goldsmith, P.L. (eds) (1972)*Statistical Methods in Research and Production*, MacMillan Publishing, New York.
[46] Walden, M., Springs, W. and Mariahazey-Westchester, A. (1971) US Patent 3,625,891, assigned to Armour Industrial Company.
[47] Ackerman, J., Miller, M. and Whittlinger, D. (1993) US Patent 5,221,794, assigned to Sherex Chemical Co. Inc.
[48] Brock, M. Enneking, M. and Kosswig, K. (1996) European Patent Application 718,275, assigned to Huels AG.
[49] J. Bigorra Llosas, *et al.* (1996) German Patent Application DE 4439090, assigned to Henkel.
[50] Haq, Z. *et al.* (1995) W.O. Patent Application 95/27771, assigned to Unilever.
[51] Straathof, T. and A. Konig (1988) US Patent 4,767,547, assigned to Procter and Gamble.
[52] Baker, E. (1994) *Preparation, Properties and Formulation of DEEDMAC: An Environmentally Friendly Cationic Surfactant*, oral presentation at the 85th Annual AOCS National Meeting, Atlanta, GA.
[53] Ellis, S.R. and Turner, G.A. (1996) US Patent 5,516,437, assigned to Lever Brothers.
[54] Wilsch-Irrgang, A. *et al.* (1996) German Patent Application DE 4,436,714, assigned to Henkel.
[55] Grandmaire, J.P. and Hermosilla, A. (1996) US Patent 5,525,245, assigned to Colgate Palmolive.
[56] Swartley, D.M. *et al.* (1995) US Patent 5,399,272, assigned to Procter and Gamble.
[57] Corona, A. *et al.* (1996) European Patent Application 704,522, assigned to Procter and Gamble.
[58] Lam, A.C. *et al.* (1996) US Patent 5,480,567, assigned to Unilever.
[59] Morita, H. and Oota, S. (1995) Japanese Patent Application, 07018578, assigned to Lion.
[60] Iacobucci, P. (1997) *Tumble Dryer Sheets*, oral presentation at the 88th AOCS National Meeting, Seattle, WA.
[61] Puchta, R. *et al.* (1994) W.O. Patent 94/02676, assigned to Henkel.
[62] De Buzzaccarini, F., Farwick, T.J. and Zhen, Y. (1995) W.O. Patent 95/29218, assigned to Procter and Gamble.
[63] De Buzzaccarini, F. Farwick, T.J. and Zhen, Y. (1995) W.O. Patent, 95/29217, assigned to Procter and Gamble.
[64] Vick, H. (1991) W.O. Patent 5,019,376, assigned to S.C. Johnson.
[65] Deshpande, V. and Walts, J. (1991) US Patent 5,034,219, assigned to Sterling Drug.
[66] Nakama, Y. *et al.* (1990) US Patent 4, 9119,846, assigned to Shiseido Co. Ltd.
[67] Pings, K.D. (1996) US Patent 5,482,703, assigned to Procter and Gamble.
[68] Pings, K.D. (1997) US Patent 5,482,703, assigned to Procter and Gamble.
[69] Robbins, C.R. and. Patel, A.M (1997) US Patent 5,415,857 assigned to Colgate Palmolive.
[70] Akzo Nobel Chemicals, *Akzo Nobel Personal Care: Product Guide*, Akzo Nobel Chemicals Inc., Chicago, IL, 1996.
[71] Murphy, A.P. (1981) US Patent 4,259,217, assigned to Procter and Gamble.
[72] Mehreteab, A. and Loprest, F.J. (1995) US Patent 5,441,541 assigned to Colgate Palmolive.
[73] Mehreteab, A. and Loprest, F.J. (1995) US Patent 5,472,455 assigned to Colgate Palmolive.
[74] Rorig, H. and Porta, N. (1992) US Patent 5,078,896, assigned to Akzo N.V.
[75] Smith, W.L. (1991) US Patent 5,055,219, assigned to the Clorox Company.
[76] Leach, M.J. (1995) W.O. Patent 96/26257, assigned to Unilever.
[77] Candau, F. (1992) in *Polymerization in Organized Media* (ed. M. Paleos), Gordon & Breach, p. 215
[78] Chieng, T. Gan, L., Chew, C., Ng, S. and Pey, K. (1996) *Polymer*, **37** (13) 2801.
[79] Chieng, T. Gan, L., Chew, C., Ng, S. and Pey, K. (1996) *Polymer*, **37** (21) 4823.
[80] Escalante, J., Rodriguez-Guadarrama, L., Mendizabal, E., Puig, J., Lopez R. and Katime, I. (1996) *Journal of Applied Polymer Science*, **62** (9) 1313.
[81] Santhanalakshmi, J. and Anandhi, K. (1996) *Colloid and Polymer Science*, **274** (8) 772.

[82] Vavak, V. Pavlov, A., Svitova, T., Danilenko, A., Egorov, V. and Varlamova, E. (1996) *Colloid Journal* (translation of *Kolliodn. Zh.*), **58** (1) 1
[83] Egorov, V., Suprun N. and Klyamkin, A. (1996) *Vestn. Mosk. Univ., Ser. 2, Khim.*, **37** (2) 178.
[84] Mizutari, T. *et al.* (1996) Japanese Patent 07,185,296, assigned to Ashahi Denka Kogyo Kk.
[85] Dreja, M. and Tieke, B. (1996) *Macromol. Rapid Comm.*, **17** (11) 825.
[86] Li, T., Gan, L., Chew, C., Teo, W. and Gan, L. (1996) *Langmuir,* **12** (24) 5863.
[87] Boutevin, B., Robin, J.-J., Boyer, B., Roque, J.-P. and Senhaji, O. (1996) *European Polymer Journal,* **32** (7) 821.
[88] McGrath, K. (1996) *Colloid and Polymer Science*, **274** (5) 399.
[89] Knudson, M. and Jones, T. (1992) US Patent 5,160,454, assigned to Southern Clay Products.
[90] Tso, Su, Beall, G. and Holthouser, M. (1987)US Patent 4,677,158, assigned to United Catalysts Inc.
[91] Ogura, Y. *et al.* (1993) US Patent 5,202,114, assigned to Shiseido Company, Ltd.
[92] Munzy, C., Butler, B., Hanley, H., Tsvetkov F. and Peiffer, D. (1996) *Mater. Lett.*, **28** (4–6) 379.
[93] Kim, Y.S., Song, D., Jeon, J. and Choi, S. (1996) *Separation Science and Technology*, **31** (20) 2815.
[94] Sun, J. and. Yu, B. (1996) *Feijinshukuang*, **1** 37.
[95] Dickey, M. and Carron, K. (1996) *Langmuir*, **12** (9) 2226.
[96] Zhang, Z., Sparks, D. and. Scrivner, N (1993) *Environmental Science and Technology*, **27** 1625.
[97] Hu, Y. and Matthys, E. (1996) *Rheologica Acta,* **35** (5) 470.
[98] Brown, M. and Burris, D. (1996) *Ground Water*, **34** (4) 734.
[99] Hayworth, J. and Burris, D. (1996) *Ground Water*, **34** (2) 274.
[100] Vanjara, A. and Dixit, S. *Separation Technology*, **6** (1) 91.
[101] Tharapiwattananon, N., Scamehorn, J., Osuwan, S. Harwell J. and Haller, K. (1996) *Separation Science and Technology*, **31** (9) 1233.
[102] Bloor, D., Mwakibete, H. and Hezron, K. (1996) *J. Coll. Interf. Sci.*, **178** (1) 334.
[103] Busserolles, K., Roux-Desgranges, G. and Roux, A. (1995) *Calorim. Anal. Therm.*, **26** 76.
[104] Faes, H., De Schryver, F., Sein, A, Bijma, K., Kevelam, J. and Engberts, J. (1996) *Macromolecule*, **29** (11) 3875.
[105] Zhang, K., Xu, B., Winnik, M. and Macdonald, P. (1996) *Journal of Physical Chemistry*, **100** (23) 9834.
[106] Jones, C.E. (1996) A Study of the Interaction of Hydrophobically-Modified Polymers with Surfactants, *Proceedings, 4th World Surfactants Congress*, Barcelona, vol. 2, p. 439.
[107] Goldzal, A., Costeux, S. and Djabourov, M. (1996) *Colloids and Surfaces A—Physicochemical and Engineering Aspects*, **112** (2–3) 141.
[108] Cabane, B., Lindell, K., Engstroem, S. and B. Lindman (1996) *Macromolecules*, **29** (9) 3188.
[109] Hansson, P. and Almgren, M. (1996) *Journal of Physical Chemistry*, **100** (21) 9038.
[110] Winnik, F., Regismond, S. and Goddard, E. (1996) *Colloids and Surfaces A*, **106** (2–3) 243.
[111] Kastner, U., Hoffmann, H., Donges, R. and Ehrler R. (1996) *Colloids and Surfaces A—Physicochemical, and Engineering Aspects*, **112** (2–3) 209.
[112] Gong, J., Mizutani, T. and Osada, Y. (1996) *Polym. Adv. Technol.*, **7** (10) 797.
[113] Vasserman, A., Zakharova, Y., Motyakin, M. and Kasaikin, V. (1996) *Colloid Journal* (translation of *Kolliodn. Zh.*), **58** (4) 431.
[114] Maltesh, C., Kiefer, J. and Somasundaran, P. (1996) *Surfactant Science Series*, **64** 179.
[115] Kabanov, V., Zezin, A., Kasaikin, V., Efremov, V. and Sergeyev, V. (1996) *Proceedings of the 87th Annual Meeting of the Air Waste Management Association*, Paper 94-WP104.03.
[116] De Oliveira, Tiera, M. and Neuman, M. (1996) *Langmuir*, **12** (3) 607.
[117] Antonietti, M. and Conrad, J. (1994) *Angew. Chem.*, **106** (18) 1927.
[118] Asnacios, A., Langevin, D. and Argillier, J. (1996) *Macromolecules*, **29** (3) 7412.

[119] Shirahama, K., Sato, S., Niino, M. and Takisawa, N. (1996) *Colloids and Surfaces A*, **112** (2–3) 233.
[120] Wang, X., Tang, J. and Jiang, L. (1996) *Chin. Chem. Lett.*, **7** (9) 875.
[121] Anthony, O. and Zana, R. (1996) *Langmuir*, **12** (8) 1967.
[122] Anthony, O. and Zana, R. (1996) *Langmuir*, **12** (15) 3590.
[123] Semchyschyn, D., Carbone, M. and Macdonald, P. (1996) *Langmuir*, **12** (2) 253.
[124] Miura, S. and Kitazawa, S. (1996) Japanese Patent 08,157,626[96,157,626], assigned to Teijin Ltd.
[125] Cipriano, R. and Longoria, J. (1996) US Patent 5,525,642, assigned to Dow Chemical Co.
[126] Schwarz, H., Apostel, R., Richau, K. and Paul, D. (1996) *Proceedings of the 7th International Conference on Perevaporation Processes in the Chemical Industry*, p. 374.
[127] Murphy, A., Hill, A. and Vincent, B. (1996) *Ber. Bunsen-Ges.*, **100** (6) 963.
[128] Tomas Eriksson, L., Claesson, P., Eriksson, J. and Yaminsky, V. (1996) *Journal of Colloid and Interface Science*, **181** (2) 476.
[129] Liu, Y. , Evans, D., Song, Q. and Grainger, D. (1996) *Langmuir*, **12** (5) 1235.
[130] Yoon, R.-H. and Ravishankar, S. (1996) *Journal of Colloid and Interface Science*, **179** (2) 403.
[131] Becerro, A., Castro, M. and Thomas, R. (1996) *Colloids and Surfaces A—Physicochemical and Engineering Aspects*, **119** (2–3) 189.
[132] Treiner, C. (1996) *Actual. Chim.* **2–3** 35.
[133] Favoriti, P., Monticone, V. and Treiner, C. (1996) *Journal of Colloid and Interface Science*, **179** (1) 173.
[134] Huang, L. and Somasundaran, P. (1996) *Colloids and Surfaces A—Physicochemical and Engineering Aspects*, **117** (3) 235.
[135] Seidel, J., Wittrock, C. and Kohler, H. (1996) *Langmuir*, **12** (23) 5557.
[136] Furst, E., Pagac, E. and Tilton, R. (1996) *Ind. Eng. Chem. Res.*, **35** (5) 1566.
[137] Esumi, K., Matoba, M. and Yamanaka, Y. (1996) *Langmuir*, **12** (9) 2130.
[138] Goloub, T., Koopal, L., Bijsterbosch, B. and Sidorova, M. (1996) *Langmuir*, **12** (13) 3188.
[139] Szekerres, M., Tombacz, E. and Dekany, I. (1996) *Olaj, Szappan, Kozmet.*, **45** (2) 70.
[140] Yaminsky, V., Jones, C., Yaminsky, F. and Ninham, B. (1996) *Langmuir*, **12** (15) 3531.
[141] Favoriti, P., Mannebach, M. and Treiner, C. (1996) *Langmuir*, **12** (20) 4691.
[142] Wang, L.-Q., Liu, J., Exarhos, G. and Bunker, B. (1996) *Langmuir*, **12** (11) 2663.
[143] Kitiyanan, B., O'Haver, J., Harwell, J. and Osuwan, S. (1996) *Langmuir*, **12** (9) 2162.
[144] Lee, Y., Surjadi, D. and Rathman, J. (1996) *Langmuir*, **12** (26) 6202.
[145] Wang, L.-Q., Liu, J., Exarhos, G. and Bunker, B. (1996) *Langmuir*, **12** (11) 2663.
[146] Moulik, S., Haque, M., Jana P. and Das, A. (1996) *Journal of Physical Chemistry*, **100** (2) 701.
[147] Blandamer, M., Cullis, P., Soldi, L., Rao, C. and Subha, M. (1996) *J. Therm. Anal.*, **46** (6) 1583.
[148] Kacperskam, A. (1995) *J. Therm. Anal.*, **45** (4) 703.
[149] Haas, S. and Hoffmann, H. (1996) *Progr. Coll. Polym. Sci.*, **101** 131.
[150] Eastoe, J., Dong, J., Hetherington, K., Steytler, D. and Heenan, R. (1996) *Journal of the Chemical Society, Faraday Transactions*, **92** (1) 65.
[151] Yasushi, H., Mababe, M., Nomura, M., Inque, T., Murata, Y. and Sasaki, Y. (1996) *Nippon Kagaku Kaishi*, **10** 861.
[152] Zabaloy, M. and Vera, J., (1996) *J. Chem. & Eng. Data*, **41** (6) 1499.
[153] Rosen, M. and Zhu, Y.-P. (1996) *J. Chem. Eng. Data*, **41** 1160
[154] Van Ginkel, C. (1995) Biodegradability of Cationic Surfactants, in *Biodegradable Surfactants* (eds. D. Karsa and M. Porter), Blackie Academic & Professional, Chapman & Hall, London, p. 183.
[155] Garcia, M., Ribosa, I., Leal, J. and Vives, J. (1996) Effects of the Hydrophobic Moiety of Quaternary Ammonium Surfactants on their Ecological Properties, in *Proceedings of the 4th World Surfactants Congress*, Barcelona, vol. 4, p. 373.

[156] Kim, T.-S., Hirao, T. and Ikeda, I. (1996) *Journal of the American Oil Chemists' Society*, **73** (1) 67.
[157] Kim, T.-S., Kida, T., Nakatsuji, Y., Hirao, T. and Ikeda, I. (1996) *Journal of the American Oil Chemists' Society*, **73** (7) 907.
[158] Hirata, H., Hattori, N., Ishida, M., Okabayashi, H, Hirofumi, F. and Zana, R. (1995) *Journal of Physical Chemistry*, **99** (50) 17778.
[159] De, S., Aswal, V., Goyal, P. and Bhattacharya, S. (1996) *Journal of Physical Chemistry*, **100** (28), 11664.
[160] Rosen, M. and Liu, L. (1996) *Journal of the American Oil Chemists' Society*, **73** (7) 885.
[161] Dam, T., Engberts, J., Karhauser, J., Karaborni, S. and van Os, N. (1996) *Colloids and Surfaces, A*, **118** (12) 41.
[162] Esumi, K., Goino, M. and Koide, Y. (1996) *Journal of Colloid and Interface Science*, **183** (2) 539.
[163] Esumi, K., Taguma, K. and Koide, Y. (1996) *Langmuir*, **12** (16) 4039.
[164] Perez, L., Ribosa, I., Garcia, T., Manresa, A. and Infante, M.R. (1996) Synthesis and Properties of Gemini Cationic Surfactants from Arginine, in *Proceedings of the 4th World Surfactants Congress*, Barcelona, vol. 1, p. 559
[165] Zana, R. (1996) *Colloid and Interface Science*, **1** (5) 556.
[166] Li, X., Lin, E., Zhao, G. and Txiao (1996) *Journal of Colloid and Interface Science*, **184** (1) 20.
[167] Li, X., Zhang, G., Zhang, M. and Wu, G. *Colloids and Surfaces, A*, **111** (3) 161.
[168] Gao, Y. and Zheng, Y. (1996) *Acta Chim. Sinica*, **45** (5) 491.
[169] Huang J. and Zhao, G. (1996) *Colloid Polym. Sci.*, **274** (8) 747.
[170] Brasher, L., Herrington, K. and Kaler, E. (1996) *Langmuir*, **11** (11) 4267.
[171] Rodriguez, C. and Scamehorn, J. (1996) Kinetics of Surfactant Precipitation, in *Proceedings of the 4th World Surfactants Congress*, Barcelona, vol. 2, p. 188.
[172] Bergstrom, M. (1996) *Langmuir*, **12** (10) 2454.
[173] Haque, Das, A., Rakshit, A. and Moulik, S. (1996) *Langmuir*, **12** (17) 4084.
[174] Regev, O. and Khan, A. (1996) *Journal of Colloid and Interface Science*, **182** (1) 95.
[175] Caria, A. and Khan, A. (1996) *Langmuir*, **12** (26) 6282.
[176] Blandamer, Briggs, B., Cullis, P., Engberts, J. and. Kacperska, A (1996) *Journal of the Chemical Society, Faraday Transactions*, **91** (23) 4275.
[177] Liu, L. and Rosen, M. (1996) *Journal of Colloid and Interface Science*, **179** (2) 454.
[178] Esumi, Goino, M. and Koide, Y. (1996) *Colloids and Surfaces A—Physicochemical and Engineering Aspects*, **118** (1–2) 161.
[179] Penfold, J., Staples, E., Cummins, P., Tucker, I., Thompson, L., Simister, E. and Lu, J. (1996) *Journal of the Chemical Society, Faraday Transactions*, **92** (10) 1773.
[180] Penfold, J., Staples, E., Cummins, P., Tucker, I., Thomas, R., Simister, E. and Lu, J. (1996) *Journal of the Chemical Society, Faraday Transactions*, **92** (9) 1549.
[181] Huang, L. and Somasundaran, P. (1996) *Langmuir*, **12** (24) 5790.
[182] Desai, T. and Dixit, S. (1996) *Journal of Colloid and Interface Science*, **179** (2) 544.
[183] Zoeller, N., Shiloach, A. and Blankschtein, D. (1996) *CHEMTECH*, **26** (3) 24.
[184] Boen Ho, O. (1996) The Determination of Davies HLB Increments for Cationic Surfactants from Zeta Potential Measurements on Emulsions, in *Proceedings of the 4th World Surfactants Congress*, vol. 2, p. 451.

4 Biodetergents

Tomiko Fujii

4.1 Introduction

Large numbers of natural surfactants are obtained from animal, plant and microbial origins in nature. Most natural surfactants, except some products such as lecithin, saponin and bile acid, are not often utilized as industrial resources since the amounts available are very small, and isolation and purification are rather difficult.

Extensive research and development on the fermentation of hydrocarbons since the last half of the 1960s has shown that microorganisms produce abundant extracellular surface-active substances. Biosurfactants are sometimes called microbial surfactants because they are produced and accumulated extracellularly by microorganisms such as bacteria, yeast and fungi.

More recently, large numbers of different chemical types produced by a wide variety of microorganisms have been found and the important role of these biosurfactants has been clarified. Hydrocarbon-degrading microorganisms produce extracellular surface-active substances, emulsify hydrocarbons and stimulate growth on hydrocarbons. In addition, much research has been reported on biological functions of biosurfactants, for example their presence as major components of microbial membranes, their function as the controlling materials of cell membrane structure, their behavior as carriers, their role in antibiosis as a means of survival, their antibacterial and antiviral properties, their function as receptors for bacteriophage and their role in energy storage [1, 2].

Unlike a chemically synthesized surfactant, a biosurfactant has not only a unique chemical structure but also special properties, such as a strong surface activity, high biodegradability, low toxicity and biological activity. Therefore, biosurfactants are expected to be the industrial and renewable resource for the next generation when preservation of resources, environmental protection and safety will be key issues.

A number of books [3, 4] and reviews [1, 5] have been published on the production, structure and properties, and applications of biosurfactants. This chapter surveys the different types of biosurfactant and their surfactant properties, and introduces the author's work on their applications as biodetergents and biomimetic surfactants.

4.2 Types of biosurfactant and their properties

Some microbial surface-active substances can be easily isolated by means of liquid chromatography or thin layer chromatography. The chemical structures of the pure chemicals can be identified by IR, NMR and MS. Furthermore, by clarifying the relationship between the chemical structures and the functions of the biosurfactants, intensive research has been targeted on chemical modification and production of derivatives to give more effective and more functional properties and design more biomimetic functional substances.

The hydrophobic group of biosurfactants is the long hydrocarbon chain of fatty acid and hydroxyl fatty acid while the hydrophilic group is either a saccharide, a carboxylate group, a phosphate group, a glycoprotein or a cyclic peptide [6]. Biosurfactants are usually classified in five types according to the nature of the hydrophilic group (Table 4.1).

Table 4.1 Biosurfactants produced by microorganisms [2] (reproduced with permission from *Journal of the Japan Oil Chemists' Society*, **41** (1992), published by the Japan Oil Chemists' Society

Biosurfactants	Producing organisms
Glycolipids	
Rhamnolipids	*Pseudomonas* sp.
Trehalose lipids	*Rordococcus* sp.
	Arthrobacter sp.
	Mycobacterium sp.
Sophorolipids	*Torulopsis* sp.
Glucose lipids	*Corynebacterium* sp.
Cellobiose lipids	*Ustilago* sp.
Oligosaccharide lipids	*Rhodococcus* sp.
Mannosylerythritol lipids	*Candida* sp.
Fatty acids	
Corynomycolic acids	*Corynebacterium* sp.
Mycolic acids	*Mycobacterium* sp.
Spiculisporic acids	*Penicillium* sp.
Agaricic acids	*Polyporus* sp.
Phospholipids	*Thiobacillus* sp.
	Corynebacterium sp.
Polymers	
Emulsan	*Acinetobacter* sp.
Liposan	*Candida* sp.
Glycoprotein	*Pseudomonas* sp.
Polysaccharide–protein complex	*Corynebacterium* sp.
Lipopeptides	
Surfactin	*Bacillus subtilis* sp.
Viscosin	*Pseudomonas* sp.
Ornithine lipids	*Pseudomonas* sp.
N-acyl amino acids	*Crustaceans*

Most biosurfactants have either non-ionic or anionic character. Cationic biosurfactants include the antibiotics gramicidin S and polymyxin B.

4.2.1 *Glycolipids*

Since glycolipids can be produced from microorganisms in relatively high yields, commercial production and applications are possible. The representative products are rhamnolipids, trehalose lipids and sophorolipids.

4.2.1.1 Rhamnolipids. Six rhamnolipid homologues are known, varying in the numbers of rhamnose and β-hydroxydecanoic acid units present (Figure 4.1).

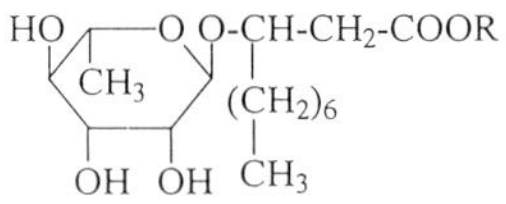

Rhamnolipid R1 : R = -CH-CH$_2$COOH | (CH$_2$)$_6$ | CH$_3$

Rhamnolipid R2 : R = H

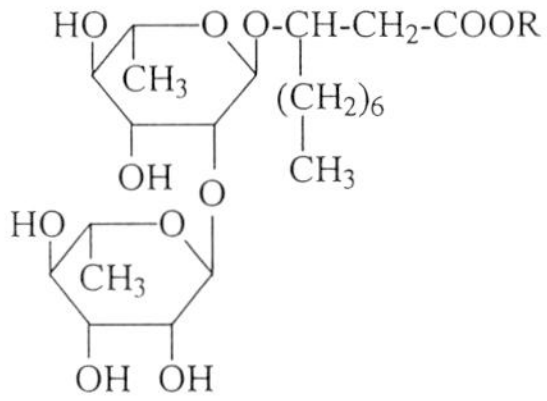

Rhamnolipid R3 : R = -CH-CH$_2$COOH | (CH$_2$)$_6$ | CH$_3$

Rhamnolipid R4 : R = H

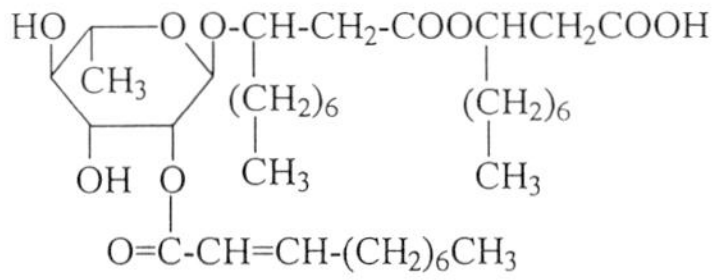

Rhamnolipid R5

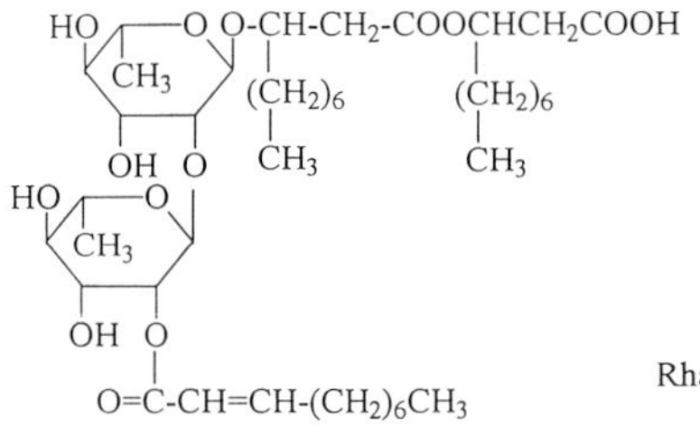

Rhamnolipid R6

The minimum surface tension of aqueous solutions of rhamnolipids R1 and R3 is 25 mN/m. The interfacial tension of R3 against kerosene is 0.2 mN/m and that of R1 is 1.0 mN/m [7]. The minimum surface tension of the sodium salts of rhamnolipids R5 and R6 is 28 mN/m. The interfacial tensions of R5 and R6 against kerosene are 0.2 mN/m and 3.2 mN/m, respectively [8]. The methyl ester of rhamnolipid R6 lowers the interfacial tension considerably at the octane–solution interface (0.1 mN/m), shows an excellent wetting action on the surface of biomembranes and synthetic polymers, and is not harmful to biomembranes [9]. As rhamnolipids R5 and R6 each have two hydrophilic groups, one saccharide and one carboxylic acid group, the morphologies of their molecular aggregates reversibly change from vesicles under acidic conditions (at pH 4.3–5.8) to lamella (at pH 6.0–6.5), lipid particles (at pH 6.2–6.6), and finally to micelles (at pH > 6.8), depending on environmental pH values [10]. The pyrene derivative of rhamnolipid is used as a surface-active fluorescent probe in the estimation of the polarity and fluidity of colloidal interfaces and biosurfaces [11].

Due to their excellent surface activity, rhamnolipids may be used as emulsifiers, detergents, stimulating agents for growth of hydrocarbon-degrading microorganisms, activating agents in waste water containing hydrocarbon and in enhanced oil recovery. The enhanced removal of oil from gravel needs large quantities of economical surfactants. At pilot-plant scale, final product concentrations of *Pseudomonas* sp. can reach 2.0 g/l [12].

Glucose lipids, which have chemical structures similar to rhamnolipids, exhibit high surface activity in spite of their molecular bulkiness. The minimum surface and interfacial tensions of glucose lipid against hexadecane are 24.6 and 13.3 mN/m, respectively [13].

4.2.1.2 Trehalose lipids. Trehalose lipids are found in the culture broth of *Arthrobacter paraffineus*, *Corynebacterium* sp., *Mycobacterium* sp. and *Rhodococcus erythropolis* (Figure 4.2).

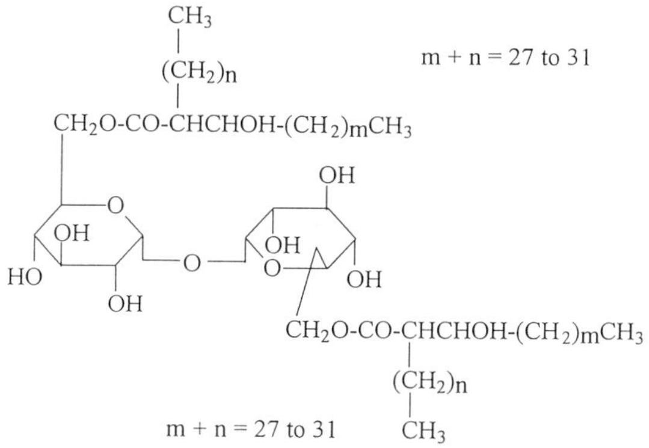

Figure 4.2 Trehalose lipids (trehalose-6-6'dicorynomycolate) from *Rhodococcus erythropolis*.

The minimum surface tensions of trehalose-6-corynomycolate, trehalose-6,6'-dicorynomycolate and trehalose-2,3,4,2'-tetraester from *R. erythropolis* are 32, 36 and 26 mN/m, respectively, in synthetic hard water. The interfacial tensions against *n*-hexadecane are 16, 17 and <1 mN/m, respectively [6]. Trehalose lipids are chemically stable and their surface activities are independent over a wide range of temperature, pH values and salt concentrations. Because of their excellent surface activity, which is similar to that of rhamnolipids, trehalose lipids are utilized as emulsifiers, detergents and stimulating agents for the growth of hydrocarbon-degrading microorganisms. They are also used for the tertiary recovery of petroleum from sandstone and to diminish the effects of oil pollution.

4.2.1.3 Sophorolipids. Sophorolipids are produced on a relatively large scale from *Tolulopsis bombicola* in the culture broth using glucose as the carbon source. Sophorolipids consist of more than nine kinds of homologues as well as both the free acid and lactone forms of the terminal carboxylic moiety [14, 15]. Sophorolipids are not used as emulsifiers since they are not effective at stabilizing water-in-oil emulsions. Sophorolipids are converted into the unitary glycolipids by modifying the stable glycosyl ether bond with acid or alkali and are subsequently derived to various compounds having a different hydrophilic–lipophilic balance. In particular, the propyleneglycol adduct is commercialized as a skin moisturizer and softener in cosmetics [16]. (Figure 4.3 and Table 4.2).

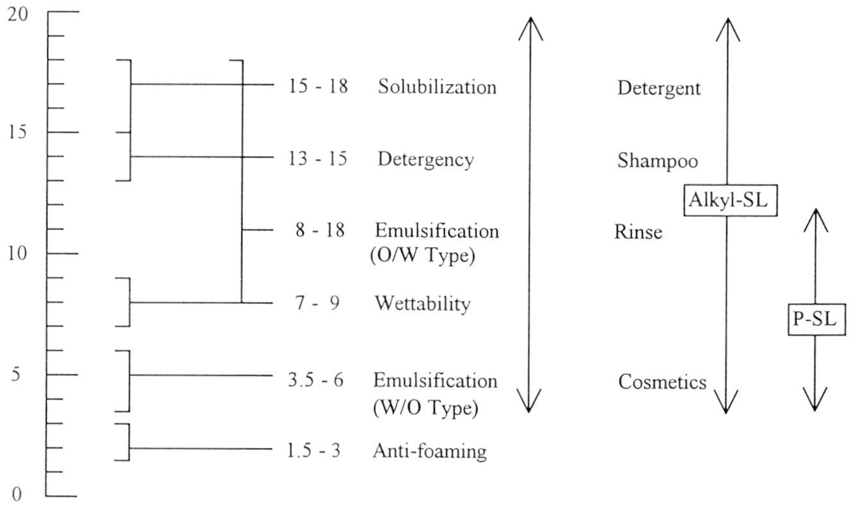

Figure 4.3 Application field of alkyl-sophorolipids [16]. Reproduced with permission from *Journal of the Japan Oil Chemists' Society*, **36** (1987), published by the Japan Oil Chemists' Society.

Table 4.2 Surface activity of sophorolipid derivatives [16] (reproduced with permission from *Journal of the Japan Oil Chemists' Society*, **36** (1987), published by the Japan Oil Chemists' Society

Alkyl-sophorolipid	Surface tension (mN/m)	CMC (wt%)	Hydrophilic–lipophilic balance
Acid-SL	34.8	3.2×10^{-2}	>25–45
Methyl-SL	38.0	5.4×10^{-3}	>25–45
Ethyl-SL	40.0	2.9×10^{-3}	≥20–45
Propyl-SL	39.5	1.4×10^{-3}	19.2
Butyl-SL	37.5	7.6×10^{-4}	18.5
Amyl-SL	36.5	4.6×10^{-4}	17.7
Hexyl-Sl	36.5	3.2×10^{-4}	17.0
Octyl-Sl	35.8	2.8×10^{-4}	15.5
Decyl-SL	38.5	5.3×10^{-4}	14.0
Myristyl-SL	48.8	6.9×10^{-3}	11.0
Oleyl-SL	—	—	7–8

4.2.2 Fatty acids

4.2.2.1 Spiculisporic acid. Spiculisporic acid is produced on a relatively large scale by *Penicillium spiculisporum* using glucose as the carbon source (Figure 4.4) [17].

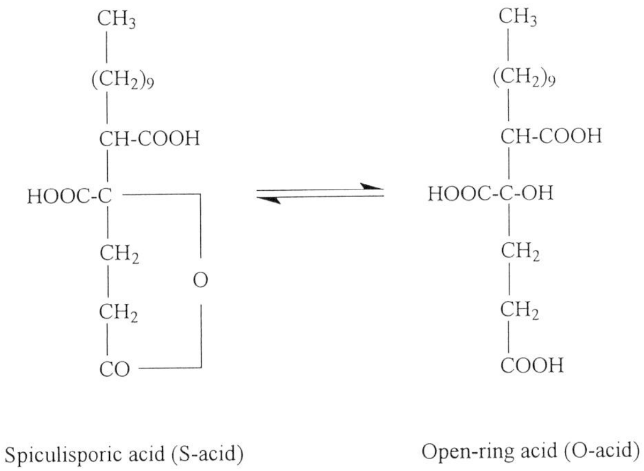

Figure 4.4 Spiculisporic acid and its ring-open acid

The sodium salts of spiculisporic acid show excellent surface activity and low foaming properties [17]. As spiculisporic acid is a polyfunctional compound with a five-membered lactone and two carboxylic groups, various derivatives can be prepared from it [18–21]. Alkylamine salts are good dispersants for α-copper phthalocyanine blue, magnetic ferric oxide and titanium dioxide (anatase) as well as good emulsifiers [22]. In addition, alkylamine salts of spiculisporic acid form vesicles in water, depending on pH [23].

4.2.2.2 Agaricic acid. Agaricic acid is contained in the body of *Polyporus officinalis Fris*. This compound is known as an antisweat pharmaceutical.

4.2.2.3 Corynomycolic acids. *Corynebacteria* produce corynomycolic acid homologues with 28 to 36 carbon atoms. Corynomycolic acids show strong surface activities over a wide range of pH values. However, the critical micelle concentration (CMC) is not well defined because of the heterogeneity of the alkyl chain lengths [24]. It is thought that the long chain of this fatty acid will yield new properties.

4.2.3 Phospholipids

Phospholipids are produced by *Thiobacillus thiooxidans* and *Corynebacterium alkanolyticum*. As phospholipids are produced on a large scale from soya lecithin, the development of the microbial phospholipid produced by microorganisms is not necessary.

4.2.4 Biopolymers

Emulsan, liposan, mannoprotein and the polysaccharide–protein complex are classified in this category. Emulsan is the lipopolysaccharide produced by *Acinetobacter calcoaceticus*. Emulsan does not appreciably reduce interfacial tension, but it is a very effective emulsifier for hydrocarbon in water, even at concentrations as low as 0.001–0.01%. It is one of the most powerful emulsion stabilizers known today. Liposan is produced by *Candida lipolytica* and consists of 83% carbohydrate and 17% protein. It is a water-soluble emulsifier and stabilizes many commercial vegetable-oil-in-water emulsions.

4.2.5 Lipopeptides

Surfactin produced by *Bacillus subtilis* is a cyclolipopeptide biosurfactant in which seven amino acids are bound to β-hydroxymyristic acid. As the cycloheptapeptide ring forms the β-sheet structure it tends to form large rod-shaped micelles and lowers the surface tension to 30 mN/m at pH 8.7 [25]. Linear heptapeptide surfactin easily forms micelles in aqueous solutions by coordinating β-sheet formation above the CMC from α-helical monomolecules below the CMC [26]. As surfactin contains both a lipophilic alkyl chain and a hydrophilic peptide chain, it forms and stabilize foams as well as bovine serum albumin

(BSA) [27]. Surfactin is sold by Calbiochem Corp. (San Diego, USA) for use in biochemical research because of its ability to inhibit blood coagulation and protein denaturation, and to accelerate fibrinolysis [28].

4.3 Properties of biodetergents

As the special properties of the various biosurfactants become known, interest in their applications, from enhanced oil recovery (EOR), the main application, to several other fields, including fine chemicals, pharmaceuticals, agrochemicals, biochemicals, cosmetics and food additives, has increased (Table 4.3).

Table 4.3 Application of biosurfactants to various industrial fields [2] reproduced with permission from *Journal of the Japan Oil Chemists' Society*, **41** (1992), published by the Japan Oil Chemists' Society).

Application fields	Biosurfactants
Fine chemicals	
Emulsifying, spreading agents	Trehalose lipids
Wetting. penetrating agents	Corynomycolic acids
Special pigments	Spiculisporic acids
Antistatic agents	
Polymer materials	
Pharmaceuticals	
Immune activation	Cord factors
Thrombus solubilization	Surfactin
Gall stone solubilization	Bile acids
Antibiotics	Sophorolipids
Drug carrier	Rhamnolipids
Agrochemicals	
Pesticides (spray applications)	Lecithin
Biochemicals	
Protein solubilizers	Sucrose lipids
Liposome materials	Phospholipids
Cosmetics	
Skin moisturizer	Sophorolipids
Skin care	Emulsan
Sweat control	Agaricic acids
Food additives	
Foaming agents	Saponin
Emulsifying agents	Lecithin
Stabilizers (bakery)	Mannosylerythritol lipids

While fatty acid soap, a natural surfactant, has excellent detergency performance, biodegradability and safety and low irritation, it also has some defects, such as low solubility and high sensitivity to hard water. Hence, the detergency performances of spiculisporic acid [29] and agaricic acid [30], a fatty acid type-biosurfactant, were studied in order to compare them with those of commercially available synthetic detergents and fatty acid soap, to evaluate them as possible alternatives to fatty acid soap. Detergents in general are required to have well-balanced properties of penetration, dispersion, emulsification and solubilization.

4.3.1 Spiculisporic acid

The surface tension of the monosodium salt (S-1Na) of spiculisporic acid (4,5-dicarboxy-4-pentadecanolide; S-acid) (Figure 4.4) decreases as the concentration increases and becomes constant at the CMC. The surface tension of S-1Na at the CMC is 34 mN/m and is lower than that of the disodium salt (S-2Na) (39 mN/m). S-1Na shows a surface activity comparable to those of sodium laurate ($C_{12}Na$) (36 mN/m), sodium dodecylbenzenesulfonate (LAS) (30 mN/m) and hexaethyleneglycol monodecylether ($C_{10}EO_6$) (33 mN/m) (Figure 4.5).

The interfacial tension of S-1Na against decane is 4.5 mN/m, which is less than that of S-2Na (10.6 mN/m), while the interfacial tensions of $C_{12}Na$, LAS and $C_{10}EO_6$ are 10.4, 1.4 and 7.1 mN/m, respectively. Removal of trioleoyl

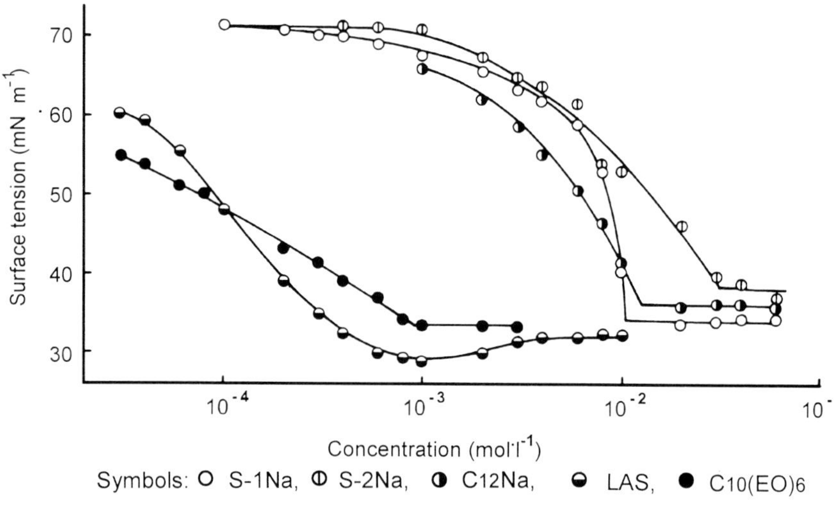

Figure 4.5 Surface tension versus concentration curves of sodium salts of spiculisporic acid and other surfactants at 25°C.

glycerol by S-acid from cotton fabric is increased markedly in the vicinity of its CMC. The removal of trioleoylglycerol by S-1Na exceeds that of S-2Na and is essentially the same as those of synthetic detergents and soap (Figure 4.6).

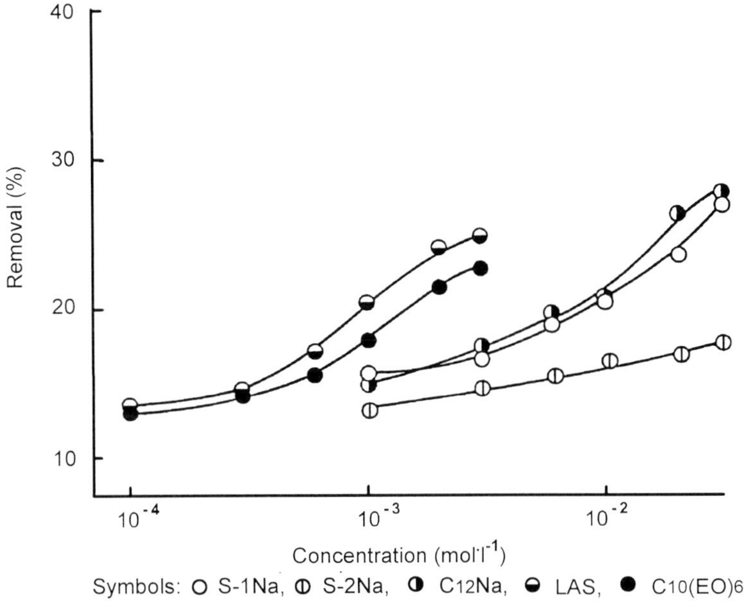

Figure 4.6 Removal of trioleoylglycerol by sodium salts of spiculisporic acid and other surfactants at 40°C.

Since S-acid is extremely hydrophilic, because of its short alkyl chain length and polycarboxylic acid, the various alkylamine salts of S-acid and the corresponding open-ring acid of the analogous compound 3-hydroxy-1,3,4-tetradecanetricarboxylic acid (O-acid) can be prepared [18]. Table 4.4 shows the removal efficiency of trioleoylglycerol by the *n*-hexylamine, cyclohexylamine, *n*-heptylamine and ethylhexylamine salts of S-acid and O-acid.

The di-neutralized *n*-hexylamine salts of S-acid and O-acid were found to have the greatest ability for soil removal. The soil removal ability of these detergents was shown to be closely correlated to interfacial tension at the oil–water interface. The detergent with the lowest interfacial tension had the best detergency performance.

4.3.2 Agaricic acid

Agaricic acid (2-hydroxy-1,2,3-nonadecanetricarboxylic acid; A-acid) is expected to have calcium-binding capacity as well as surface-active properties since it has

Table 4.4 Removal of percent trioleoylglycerol by various alkylamine salts of spiculisporic acid.

Surfactants	Concentration (%)	
	1×10^{-2}	2×10^{-2}
S-1 sodium salt	20.4	23.5
S-2 sodium salt	16.6	16.8
S-2 *n*-hexylamine salt	—	31.1
S-1 *c*-hexylamine salt	12.0	15.3
S-2 *c*-hexylamine salt	14.8	—
S-1 *n*-hepthylamine salt	21.6[a]	17.8[a]
S-1 ethylhexylamine salt	20.2[a]	14.7[a]
O-1 *n*-hexylamine salt	23.3	—
O-2 *n*-hexylamine salt	29.7	29.3
O-2 *c*-hexylamine salt	14.9	—
O-1 ethylhexylamine salt	15.7[a]	—
O-2 ethylhexylamine salt	14.5[a]	—
$C_{12}Na$	20.8	26.3
LAS[b]	20.3	24.0
$C_{10}(EO)_6$[b]	17.7	22.7

[a] Solution is not clear.
[b] Concentration ($\times10^{-3}$ M).

the chemical structure of 3-hexadecylcitric acid. The mono- (A-1Na), di- (A-2Na) and tri- (A-3Na) sodium salts of agaricic acid were prepared by neutralization with sodium methoxide in order to study their surface-active properties, calcium-binding capacities and soil-removal abilities. The surface tensions of the A-1Na, A-2Na and A-3Na salts at the CMC are 52.2, 61.0 and 53.4 mN/m, respectively. Their ability to lower the surface tension of any sodium salt of agaricic acid is less than that of synthetic detergents and soaps, and is also less than that of mono- and disodium salts of spiculisporic acid. The interfacial tensions against decane at the CMC are 19.4 mN/m for A-1Na, 27.0 mN/m for A-2Na and 18.0 mN/m for A-3Na. These values are rather higher than those of other detergents and soaps (Figure 4.7). These results show that the surface-active properties of the sodium salts of agaricic acid are somewhat inferior to those of ordinary detergents.

Agaricic acid is expected to be able to bind calcium ions because its structure contains the chemical structure of 3-hexadecylcitric acid. Therefore, the calcium-binding capacities of the sodium salts of agaricic acid were compared with that of sodium citrate. The calcium-binding capacities of the sodium salts of agaricic acid are shown in Table 4.5. A-2Na and A-3Na have calcium-binding capacities of nearly 1:1 (molar binding ratio), similar to ethylenediamine tetraacetic acid (EDTA), nitrilotriacetic acid (NTA), pentasodium tripolyphosphate (STPP) and trisodium carboxy methoxysuccinate (CMOS-3Na). It is clear that A-2Na and

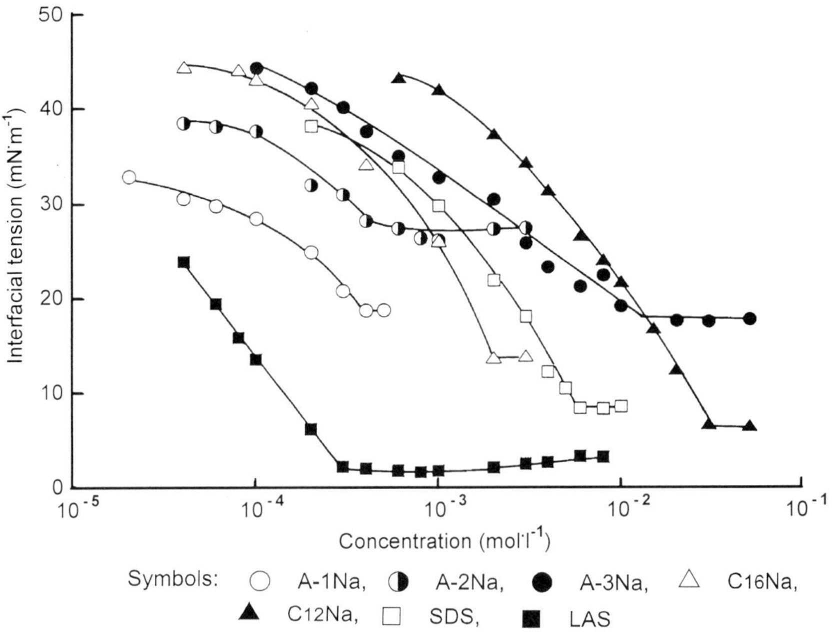

Figure 4.7 Interfacial tension at decane–water interface vs. concentration curves of sodium salts of agaricic acid and other surfactants at 25°C.

A-3Na, as well as the A-acid alkaline solution, have calcium-binding capacities twice that of sodium citrate, which has a similar chemical structure. The values for A-2Na and A-3Na are also much larger than that of the trisodium salt of spiculisporic acid (49 $CaCO_3$mg/g) [17].

The removal of trioleoylglycerol by aqueous solutions of the sodium salts of agaricic acid in distilled water and 100 ppm hard water is shown in Figures 4.8 and 4.9, respectively, in comparison with the values for other detergents. The removal ability of the sodium salts of agaricic acid in distilled water is lower than those of LAS, sodium dodecyl sulphate (SDS) and soaps. However, in 100 ppm hard water, the removal ability of the trisodium salt of agaricic acid, which has the highest calcium-binding capacity, is the same as in distilled water although those of LAS, SDS and especially soaps decrease markedly. This is because the trisodium salt of agaricic acid has an excellent calcium-binding capacity and surface-active properties.

4.3.3 Sophorolipids

As sophorolipids are mild on the skin, they may be expected to be used as detergents, shampoo bases and in cosmetics. Sophorolipids produced from

Table 4.5 Calcium-binding capacity of sodium salts of agaricic acid

		Calcium-binding capacity		
		($CaCO_3$ (mg/g)	$CaCO_3$ (M/M)	Stability constant (log *K*)
A-1 sodium salt	(pH 6)	103	0.45	1.87
A-2 sodium salt	(pH 8)	235	1.08	3.23
A-3 sodium salt	(pH 10)	255	1.23	3.21
Sodium citrate	(pH 6)	31	0.08	0.56
	(pH 8)	101	0.26	1.01
	(pH 10)	123	0.32	1.00
A-acid[a]	(pH 6)	51	0.21	0.45
	(pH 8)	212	1.85	3.01
	(pH 10)	261	1.09	2.90

[a] A-acid was dissolved in buffer solutions of pH 6, 8, and 10.

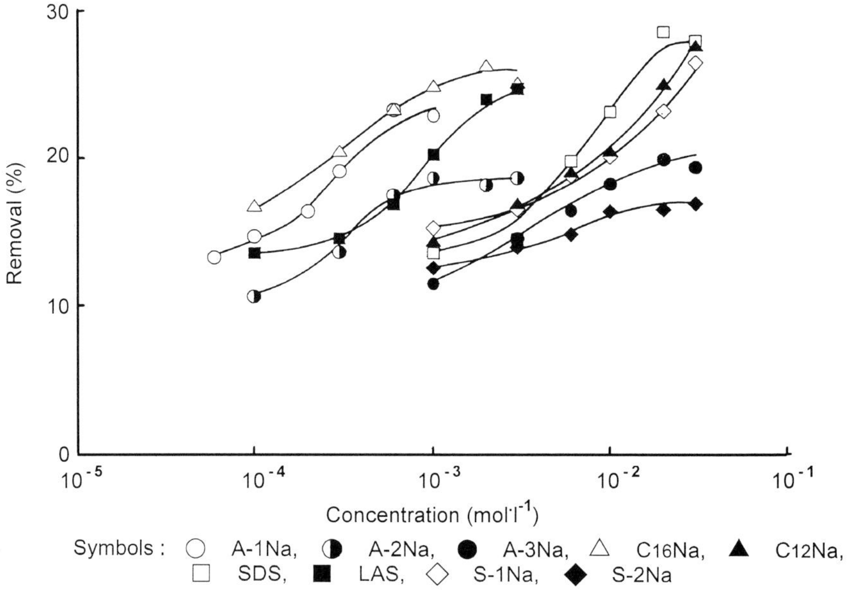

Figure 4.8 Removal of trioleylglycerol by sodium salts of agaricic acid in distilled water at 40°C.

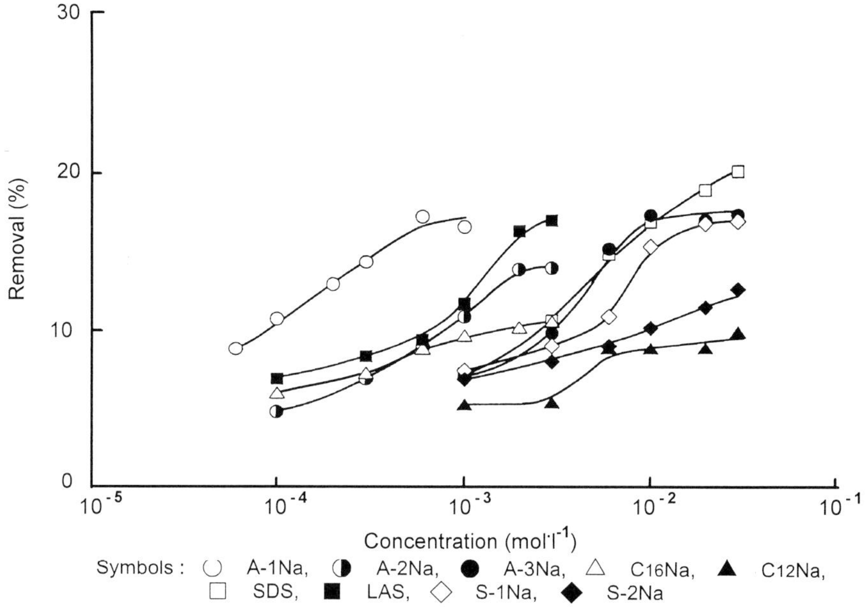

Figure 4.9 Removal of trioleoylglycerol by sodium salts of agaricic acid in 100 ppm hard water at 40°C.

Candida bombicola ATCC22214 cultured using glucose as the carbon source can be separated into six kinds of sophorolipid homologues by thin-layer chromatography–flame ionization detector (TLC-FID) analysis (Figure 4.10) [31]. The main product is sophorolipid SL-1 (42.8%). The structures of each component, SL-1 to SL-6, were identified according to reference [14] (Figure 4.11). SL-1, SL-2 and SL-3 consist of the structures of sophorose and octadecanoic acid, forming a lactone ring in which R1 and R2 in SL-1 are acetyl groups, in SL-2 are an acetyl and a hydroxyl group and in SL-3 are hydroxy groups. Formation of SL-4(a) and SL-4(b) depends on the positions available to form the lactone ring. The compounds SL-5 and SL-6 have the open acid form.

The mixture of sophorolipid homologues is neutralized with one, two or three equimolar amounts of sodium hydroxide. Figure 4.12 shows the typical IR spectra of the neutralized solution as well as the original solid and chloroform solutions. In the spectra of the solid and chloroform solutions, the absorption attributed to the C–O stretching of the lactone ring appears at $1740 cm^{-1}$. For the solutions neutralized with one (SL-I) and two (SL-II) equimolar amounts of sodium hydroxide, the absorptions at $1740 cm^{-1}$ are smaller and a new peak clearly appears at $1550 cm^{-1}$, which can be attributed to carboxylate. Therefore the lactone and acidic forms coexist in these solutions. In the case of the three (SL-Na) equimolar amounts of sodium hydroxide, the absorption appears only at $1550 cm^{-1}$ and the lactone ring is opened completely.

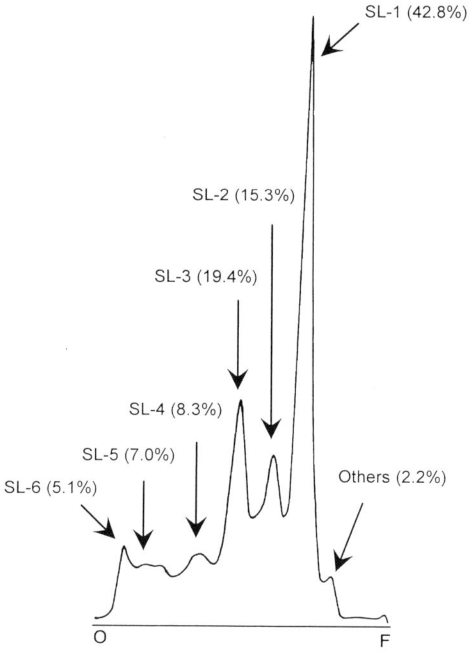

Figure 4.10 Chromatogram of sophorolipid by TLC-FID analysis. Conditions: mobile phase = $CHCl_3/CH_3OH/H_2O$ = 65/15/2; stationary phase = Chromarod S-III (latron); analyzer = latroscan TH-10 (latron).

The surface and interfacial activities of sophorolipid homologues become stronger as the lactone ring is opened. When it is completely opened (SL-Na), the interfacial tension is similar to that of SDS (Figure 4.13). Removal of trioleoylglycerol by SL-Na is the same as for dodecyl-β-D-maltoside (AM), SDS and LAS (Figure 4.14).

4.4 Synthesis and properties of biomimetic surfactants

Corynomycolic acids (2-alkyl-3-hydroxy fatty acids), biosurfactants from *Corynebacterium leps*, are highly interesting fatty acids because of their very long carbon chain lengths. However, since microbial corynomycolic acids are homologous mixtures containing different alkyl chain lengths, they are not suitable for studying surface-active and other basic properties. Therefore, corynomycolic acids with definite alkyl chain lengths were synthesized to investigate the relationship between chemical structure and surface-active properties [32].

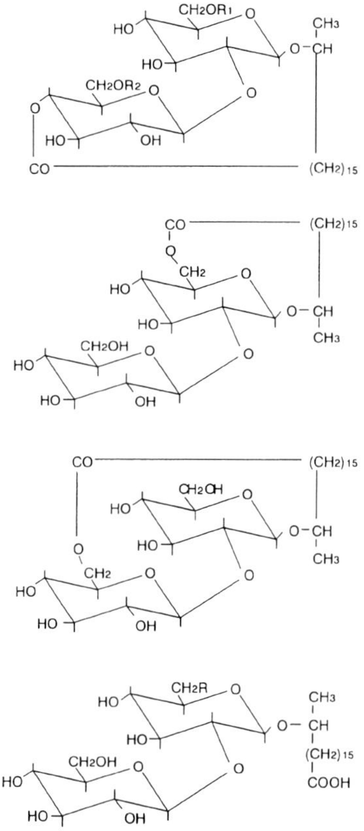

SL-1 $R_1 = R_2 = COCH_3$
SL-2 $R_1 = COCH_3$ $R_2 = H$
SL-3 $R_1 = R_2 = H$

SL-4(a)

SL-4(b)

SL-5 $R = COCH_3$
SL-6 $R = H$

Figure 4.11 Constitutional formulae of sophorolipid homologues.

Corynomycolic acids with definite alkyl chain lengths were synthesized by the method of Ishigami *et al.* [33], modified by using fatty acid ethyl esters as starting materials under the following synthetic scheme: Claisen condensation to oxo-ester, reduction to hydroxyester with sodium borohydride and hydrolysis with alkali to corynomycolic acid (Figure 4.15). In this method corynomycolic acids with definite alkyl chain lengths are relatively easily synthesized at a high yield (10–50%).

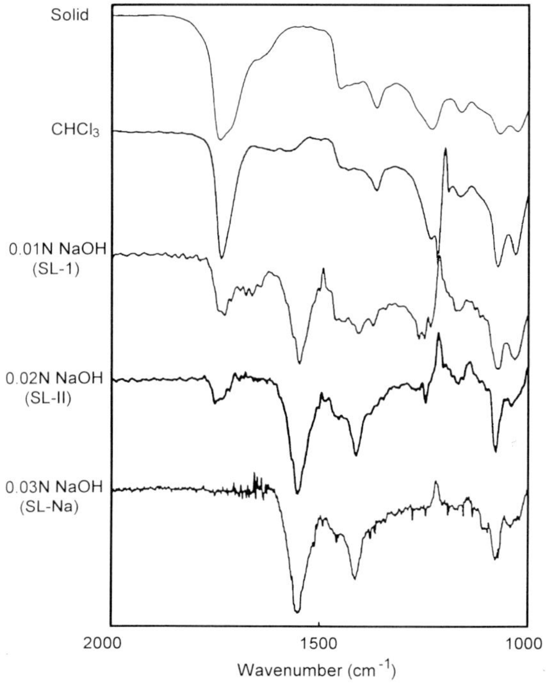

Figure 4.12 IR spectra of sophorolipid homologues taken with a Perkin Elmer 1720.

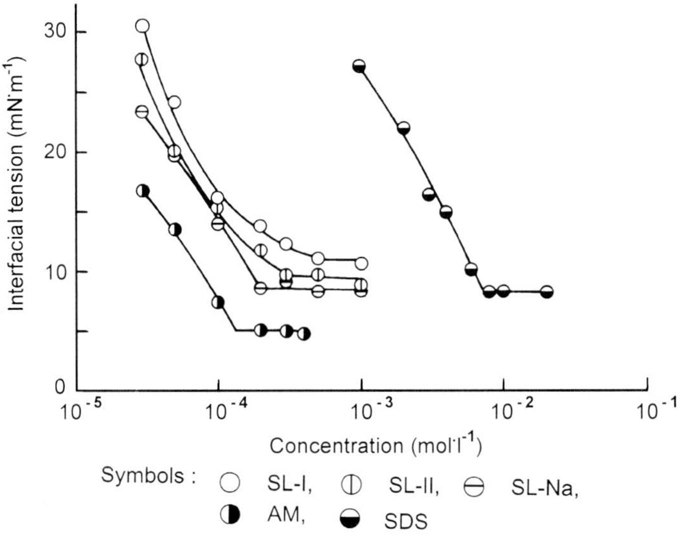

Figure 4.13 Interfacial tension at the decane–water interface versus concentration curves of sophorolipid homologues at 30°C.

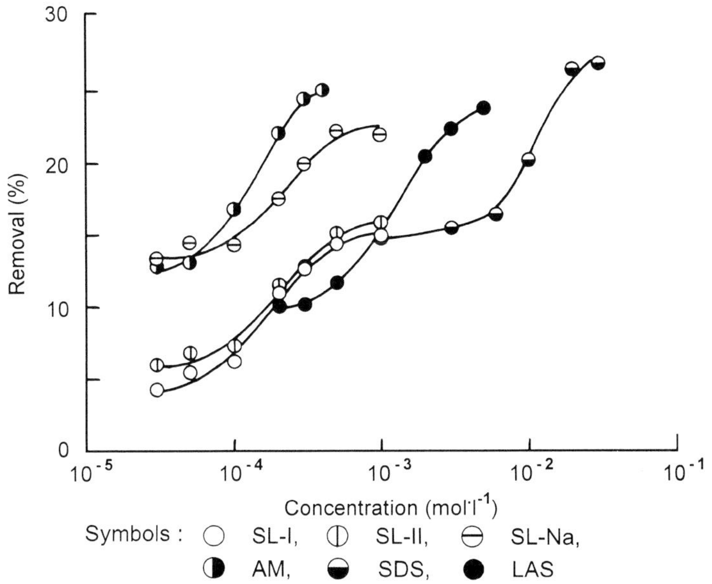

Figure 4.14 Removal of trioleoylglycerol by sophorolipid homologues at 40 °C.

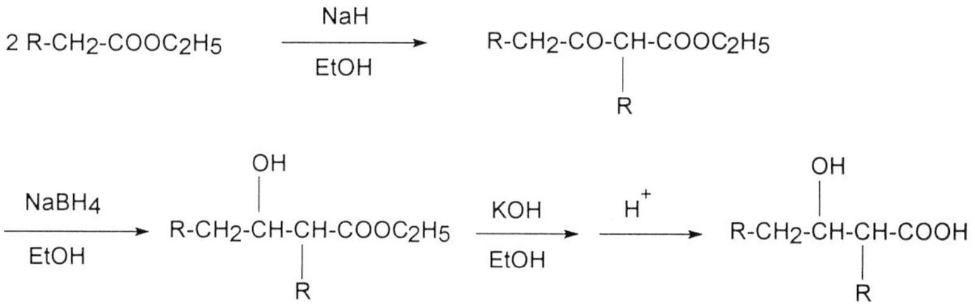

Figure 4.15 Synthesis of corynemycolic acids.

These synthetic corynomycolic acids consist of a mixture of approximately equal amounts of syn-type and anti-type isomers. The surface and interfacial tensions of synthetic corynomycolic acids decrease with increasing concentration and show a well-defined CMC, which is not seen for microbial corynomycolic acid. The values of the CMC and surface and interfacial tensions at the CMC are summarized in Table 4.6, together with the values for spiculisporic acid and agaricic acid (biosurfactants in the same category). The surface-active properties of corynomycolic acids with more than 16 carbon atoms were greater than those of monosodium salts of spiculisporic acid and agaricic acid, and were comparable to, or even better than, that of sodium LAS, a typical petrochemical surfactant. The acids with a total number of carbon atoms of 24 and 28 were insoluble in alkaline aqueous solution. Furthermore, corynomycolic acids are expected to form stable monolayers since they have the chemical structure of two lipophilic alkyl chains and two hydrophilic groups [34, 35].

Table 4.6 CMC, γ_{CMC} and interfacial tension at the decane–water interface of synthetic corynomycolic acids at 30°C

Biosurfactant	CMC (M)	γ_{CMC} (mN/m)	Interfacial tension (mN/m)
Synthetic corynomycolates			
2-butyl-3-hydroxyoctanoic acid (C_{12})	1.6×10^{-2}	43.2	29.4
2-hexyl-3-hydroxydecanoic acid (C_{16})	7.0×10^{-3}	31.2	11.4
2-octyl-3-hydroxydodecanoic acid (C_{20})	8.0×10^{-4}	25.0	1.5
Spiculisporic acid (monosodium salt)	1.2×10^{-2}	33.9	4.5
Agaricic acid (monosodium salt)	4.0×10^{-4}	52.2	19.4
Microbial corynomycolate[a]	—	40.0	10.0

[a] Reference [24].

4.5 Conclusions

Surfactants play an important part in the detergent industry. Two kinds of detergent, based on natural fatty acid soap and synthetic petrochemical-based surfactants are currently used as detergents for textile washing. However, these detergents are not always satisfactory from the point of view of resources, ecology and safety. Biosurfactants produced by microorganisms have therefore been attracting the interest as an alternative, second generation of detergent. However, for biosurfactants to replace currently used synthetic detergents, economic production and high performance will be essential.

References

[1] Rosenberg, E. (1986) *Microbial Surfactants*, CRC Critical Reviews in Biotechnology, vol.3, CRC Press, Boca Raton, pp. 109-32.

[2] Kitamoto, D. (1992) Production of biosurfactants by microorganisms. *Journal of the Japan Oil Chemists' Society*, **41** (9) 839-45.

[3] Kosaric, N., Cairns, W.L. and Gray, N.C.C. (eds) (1987) *Biosurfactants and Biotechnology*, Surfactant Science Series Vol. 25, Marcel Dekker, Inc., New York and Basel.

[4] Kosaric, N. (ed.) (1993) *Biosurfactants: Production, Properties and Applications*, Surfactant Science Series Vol. 48, Marcel Dekker, Inc., New York and Basel.

[5] Ishigami, Y. (1993) Biosurfactants face increasing interest. *Inform*, **48** (10) 1156-65.

[6] Lang, S. and Wagner, F. (1987) Structure and properties of biosurfactants, in *Biosurfactants and Biotechnology* (eds N. Kosaric, W.L. Cairns and N.C.C. Gray), Surfactant Science Series Vol. 25, Marcel Dekker, Inc., New York and Basel, pp. 21-45.

[7] Parra, J.L., Guinea, J., Manresa, M.A., Robert, M., Mercade, M.E., Comelles, F. and Bosch, M.P. (1989) Chemical characterization and physicochemical behavior of biosurfactants. *Journal of the American Oil Chemists' Society*, **66** (1) 141-45.
[8] Ishigami, Y. , Gama, Y., Yamaguchi, M., Nakahara, H. and Kamata, T. (1987) Surface active properties of rhamnolipids as microbial biosurfactants. *Journal of the Japan Oil Chemists' Society*, **36** (10) 791-96.
[9] Ishigami, Y., Gama, Y., Ishii, F. and Choi, Y.K. (1993) Colloid chemical effect of polar head moieties of a rhamnolipid-type biosurfactant. *Langmuir*, **9** (7) 1634-36.
[10] Ishigami, Y., Gama, Y., Nagahora, H., Yamaguchi, M., Nakahara, H. and Kamata, T. (1987) The pH-sensitive conversion of molecular aggregates of rhamnolipid biosurfactant. *Chemistry Letters*, **1987** 763-66.
[11] Ishigami, Y., Ishii, F., Choi, Y.K. and Kajiuchi, T. (1996) Estimation of polarity and fluidity of colloidal interfaces and biosurfaces using rhamnolipid B pyrenacylester as surface-active fluorescent probe. *Colloids and Surfaces B: Biointerfaces*, **7** 215-20.
[12] Mulligan, C.N. and Gibbs, B.F. (1993) Factors influencing the economics of biosurfactants, in *Biosurfactants: Production, Properties and Applications* (ed. N. Kosaric), Surfactant Science Series Vol. 48, Marcel Dekker, Inc., New York and Basel, pp. 329-71.
[13] Ishigami, Y., Gama, Y. , Sano, Y., Lang, S. and Wagner, F. (1994) Interfacial and micellar behavior of glucose lipids. *Biotechnology Letters*, **16** (6) 593-98.
[14] Asmer, H.J., Lang, S., Wagner, F. and Wray, V. (1988) Microbial production, structure elucidation and bioconversion of sophorose lipids, *Journal of the American Oil Chemists' Society*, **65** (9) 1460-66.
[15] Zhou, Q.H., Klekner, V. and Kosaric, N. (1992) Production of sophorose lipids by *Torulopsis bombicola* from safflower oil and glucose. *Journal of the American Oil Chemists' Society*, **69** (1) 89-91.
[16] Kimura, Y. (1987) Research and development of biosurfactants. *Journal of the Japan Oil Chemists' Society*, **36** (10) 748-53.
[17] Ishigami, Y., Yamazaki, S. and Gama, Y. (1983) Surface active properties of biosoap from spiculisporic acid. *Journal of Colloid and Interface Science*, **94** (1) 131-39.
[18] Ishigami, Y., Gama, Y., Yamazaki, S. and Suzuki, S. (1987) Characteristics of spiculisporic acid as a polycarboxylic biosurfactant, in *Proceedings of the World Conference on Biotechnology for the Fats and Oils Industry* (ed. T.H. Applewhite), American Oil Chemists' Society, Champaign, Illinois, pp. 339-42.
[19] Gama. Y., Yasumoto. M., Suzuki, H. and Ishigami, Y. (1989) Highly selective conversion of spiculisporic acid to ester, amide and imide derivatives. *Journal of the Japan Oil Chemists' Society*, **38** (4) 292-96.
[20] Choi, Y.K., Lee, C.H., Takizawa, Y., Gama, Y. and Ishigami, Y. (1993) Tricarboxylic acid biosurfactant derived from spiculisporic acid. *Journal of the Japan Oil Chemists' Society*, **42** (2) 95-99.
[21] Hong, J.J., Yang, S.M., Choi, Y.K. and Lee, C.H. (1995) Precipitation of tricarboxylic acid biosurfactant derived from spiculisporic acid with metal ions in aqueous solution. *Journal of Colloid and Interface Science*, **173** 92-103.
[22] Ishigami, Y. and Yamazaki, S. (1985) Surface active properties and application of polycarboxylic biosurfactant. *Bulletin of the National Chemical Laboratory for Industry (Kagakugijyutsu Kenkyusyo Houkoku)*, **80** 231-37.
[23] Ishigami, Y., Gama. Y. and Yamazaki, S. (1987) The pH-dependent conversion of molecular aggregation of polycarboxylic biosurfactant. *Journal of the Japan Oil Chemists' Society*, **36** (7) 490-95.
[24] Cooper, D.G., Zajic, J.E. and Denis, C. (1981) Surface active properties of a biosurfactant from *Corynebacterium lepus. Journal of the American Oil Chemists' Society*, **58** (1) 77-80.

[25] Ishigami, Y., Osman, M. and Matsumoto, H. (1995) Significance of β-sheet formation for micellization and surface adsorption of surfactin. *Colloids and Surfaces B: Biointerfaces*, **4** 341-48.

[26] Osman, M., Ishigami, Y., Ishikawa, K., Ishizuka, Y. and Holmsen, H. (1994) Dynamic transition of α-helix to β-sheet structure in linear surfactin correlating to critical micelle concentration. *Biotechnology Letters*, **16** (9) 913-18.

[27] Razafindralambo, H., Paquot, M., Baniel, A., Popineau, Y., Hbid, C., Jacques, P. and Thonart, P. (1996) Foaming properties of surfactin, a lipopeptide biosurfactant from *Bacillus subtilis*. *Journal of the American Oil Chemists' Society*, **73** (1) 149-51.

[28] Mulligan, C.N. and Gibbs, B.F. (1993) Factors influencing the economics of biosurfactants, in *Biosurfactants: Production, Properties and Applications* (ed. N. Kosaric), Surfactant Science Series Vol. 48, Marcel Dekker, Inc., New York and Basel, pp. 329-71.

[29] Fujii, T., Hashimoto, T., Yoshimura, M. and Ishigami, Y. (1990) Biodetergent. I. Detergency performance of sodium and alkylamine salts of spiculisporic acid as polycarboxylic biosurfactant. *Journal of the Japan Oil Chemists' Society*, **39** (12) 1040-44.

[30] Fujii, T., Hashimoto, T., Shimonishi, S. and Kawase, T. (1993) Biodetergent. II. Characteristics of sodium salts of agaricic acid as detergent. *Journal of the Japan Oil Chemists' Society*, **42** (7) 493-500.

[31] Fujii, T., Ozaki, Y. and Hashimoto, T. (1993) Surface active properties and detergency of sophorolipid, in *Proceedings of the Japan Research Association Annual Meeting on Textile End Uses*, Japan Research Association for Textile End Uses, Tokyo, p. 56.

[32] Fujii, T., Yuasa, R. and Kawase, T. (1995) Biodetergent. III. Synthesis and properties of corynomycolic acids as biomimetic surfactant. *Journal of the Japan Oil Chemists' Society*, **44** (6) 419-24.

[33] Ishigami, Y., Kamata, T., Gama, Y., Kaise, M., Iwahashi, H. and Someya, J. (1989) Correlation of synthetic corynomycolic acids as biosurfactant between their surface-active properties and the function of biomembranes. *Journal of the Japan Oil Chemists' Society*, **38** (12) 1001-1006.

[34] Yuasa, Y., Fujii, T. and Kawase, T. (1994) Surface active properties and monolayer of corynomycolic acids at air–water interface, in *Proceedings of the 26th Symposium on Surfactants and Detergency*, Japan Oil Chemists' Society, Tokyo, pp. 75-81.

[35] Fujii, T., Yuasa, R. and Kawase, T. (1995) Monolayer of corynomycolic acids at air–water interface, in *Proceedings of the 34th Symposium on Oil and Fat*, Japan Oil Chemists' Society, Tokyo, p. 19.

5 Applications of surfactants in personal-care products

Kazuyuki Yahagi

5.1 Introduction

Personal-care products, such as body shampoo, facial cleanser, shampoo and dentifrice, are familiar necessities of life today. People use more personal-care cleansing products, and more frequently, than in the past because their life-styles have changed and their awareness of hygiene has heightened. Moreover, there is a much greater diversity of products, in a more segmented market than ever before, in response to changing and increased consumer demand.

Personal-care products contain surfactants and use their primary function of dirt-removal. Therefore, surfactants have so far been developed in terms of cleansing ability, as well as safety, the quality of raw materials and cost reduction. However, further changes in life-style and a sense of value have recently made consumers want more than just the basic function; researchers must now offer consumers new products with new marketing concepts and therefore they have to develop materials with high functional value in order to realize this concept [1].

This chapter describes how personal-care products have been developed in Japan and examines the roles current surfactants play in the pursuit of high functional values. For these purposes body shampoo, shampoo, conditioners and skin-care products are discussed.

5.2 Body shampoo

Bar soap has been used as a body cleanser for many years. The soap is formulated chiefly with sodium salts of long-chain fatty acids; classic surfactants with simple structures. Body shampoo was introduced to the Japanese market in 1984 and expanded its share as a new type of body cleanser, competing with soap. Body shampoo is now more widely used than bar soap.

In the development of a body shampoo, importance is given to properties such as dirt-removal, the amount of foam, the feel of the foam, the residual feel of the skin and irritancy. Because the product is put directly onto skin, irritancy, along with the feel of foam and the residual feel of the skin, is considered the most important aspect for consumers when they buy a body shampoo. Various surfactants have been developed accordingly.

Salts of fatty acids, monoalkyl phosphates (MAPs) and acyl glutamate are widely used as the main anionic surfactants of a body shampoo (Table 5.1). In most cases, the fatty acid salts contain lauric, myristic, palmitic and stearic acid ions because of the high lathering ability of these salts and the favorable feel of the foam they provide; ordinarily, potassium and triethanolamine laurates are favored as the main surfactants because of their low Krafft points.

Table 5.1 Surfactants used in body shampoos

Surfactants	Feature
Soap $RCOOM$ ($M = Na$, K, TEA, NH_4)	Creamy foam Refreshing feel
Polyoxyethylene alkyl ether sulfate (ES) $C_{12}H_{25}O(CH_2CH_2O)_nSO_3M$ ($n = 2$–3; M= Na, TEA, NH_4)	Light and rough foam Difficult to rinse
Acyl isethionate (SCI) $RCOOCH_2CH_2CH_2SO_3Na$	Applicable in hard water High moist feel
Polyoxyethylene alkyl ether carboxylate (AEC) $RO(CH_2CH_2O)_nCH_2COOM$ ($n = 3$–10; M= H, Na, K, TEA)	Low irritation Applicable in hard water
Acyl glutamate (AGS) 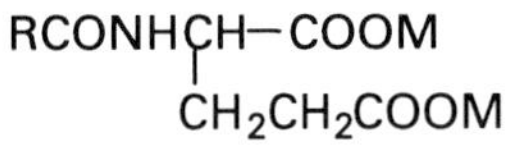	Low detergency and foaming Low irritation
Hydroxyalkyl ether carboxylate (RM2) 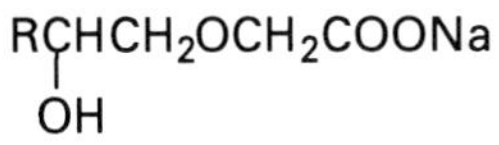	Low irritation High moist feel
Alkyl glyceryl ether sulfonate (AGES) 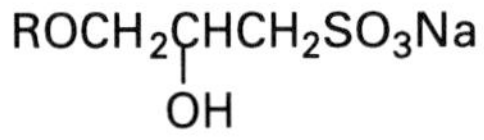	Low irritation High moist feel
Monoalkyl phosphate (MAP) 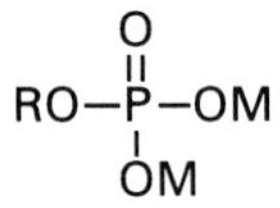 (M = Na, K, TEA)	Low irritation High foaming

It should be noted that fatty acid salts tend to form insoluble, surfactant-obstructing scum with the bivalent metallic ions, e.g. calcium and magnesium ions, often found in tap water and underground water. The salts shift the aqueous solutions towards alkalinity and may cause dermatological problems. It is advisable that a body shampoo should be formulated with auxiliary non-soap surfactants so that these adverse effects can be avoided; unlike hard bar soaps there is greater flexibility in what can be used in liquid body shampoos.

The surfactant characteristics of MAPs, their applications and dermatological actions have been reported in various papers [2, 3]. Body shampoos formulated primarily with MAPs do not dry the skin and leave a slimy feel when washed off. Moreover, because of their neutrality, they give a feel different from that of bar soaps as they are applied. These characteristics of MAP body shampoos have greatly helped to expand the body shampoo market.

When using MAPs it is also important that adverse hard-water actions should be minimized and the Krafft point adjusted so that a liquid form can be obtained. Amphoteric types of surfactant, such as laurylhydroxysulfobetaine, may be used as an auxiliary surfactant in order to solve these problems. Amphoteric surfactants form intermolecular complexes with anionic surfactants through their hydrophilic groups; the intermolecular-complex-forming action leads to the minimization of the hard-water action and an improvement in lathering. In an effort to avoid adverse effects a liquid form of an MAP potassium salt with a low Krafft point has been developed recently; this is designed to help the solution to remain liquid [4].

Acylated glutamates and their effects have been discussed in various papers [5, 6]. Of all the acylated glutamates commonly used, a triethanolamine salt, with a low Krafft point, is most likely to be found in a body shampoo formula. Consideration has also been given to other acylated amino acid salts, especially with regard to dry feel, solubility and thickening effects. Potassium-acylated glycine salts have been developed and they show a synergic effect with fatty acid salts, which improves lathering [7].

It is quite common to incorporate auxiliary surfactants into body shampoos for the purpose of improving and adjusting the foaming ability, the quality and feel of the foam, and the feel of the solution. Depending on the main surfactant used, different auxiliary surfactants are incorporated for different purposes; commonly used auxiliary surfactants are fatty acid monoethanolamide, fatty acid diethanolamide, amphoteric surfactants such as laurylhydroxysulfobetaine, alkylbetaine, alkylamidopropylbetaine and alkylimidazoline, and acylated amino acid salts such as acylated taurine.

There are other types of personal-care cleansing products besides body shampoo. These include facial cleansers (which include make-up remover) and handwash. The main and auxiliary surfactants used in these products are almost the same as those used for body shampoos.

5.3 Shampoo

It is of interest to consider the early history of hair-care products in Japan, which led up to the present diverse and detailed classification of evolved hair-care products in Japan, before discussing the application of surfactants to shampoo.

In ancient times washing hair meant religiously purging mind and body, and was seldom exercised in daily life in Japan. In the Nara (710–794) and Heian (794–1192) eras, *saikachi* (*Gleditschia japonica*) and *mukuroji* (*Sapindus mukurossi*), or soapberry, were soaked in warm water and the exudation was used for washing hair. Warm water was available from the bathtub of a temple, which Buddhism brought to Japan as this religion was introduced. It is probable that there were natural surfactants such as saponin in the exudation. A book written in the Heian era says that hair was washed on the day of *Tanabata* (July 7); it is assumed that people only washed their hair once or twice a year.

In the Edo era (1603–1867), hair was washed once or twice a month. It was done at the *engawa* (an inside corridor adjoining the outside) or *doma* (a large unfloored entrance hall) of the house or beside the *ido* (a well) of the community. Hair washing was a day-long practice; whether or not to wash hair was decided according to the weather in the morning and if the weather was good the washing solution was prepared and hair washed and dried. The process took a whole day. Illustrations of hair washing technique can be seen in *ukiyoes* (paintings or wood-block printings of the time). Oil was applied to style the hair back and clay, *udon* (thick noodle) flour, *funori* (an edible red alga), volcanic ash and ash water were used to wash it off (it is thought that the alkalinity of the wash worked on the oil and made it easy to remove).

It was not until the Meiji era (1868–1912) that soap was introduced to the market. Soap was expensive for the masses and it permeated the market slowly. However, women's hair-styles changed from the old Japanese styles that had been popular to Western styles as Western culture started to flow rapidly into Japan. This accelerated the increase in soap usage. A mixture of soap and soda ash along with other ingredients was used as a washing agent.

The term 'shampoo' was first seen and heard in the early part of the Showa era, which started in 1926; the shampoo was actually powdered soap and it was used in this form until about the 1950s. A sodium alkylsulfate (AS-Na) powder shampoo was first introduced to the Japanese market in 1955, replacing powdered soap, and shampoo entered its modern era. The development of shampoo in Japan is shown in Figure 5.1. AS-Na is better than soap in solubility and foaming power, and it only products a small amount of insoluble scum; the advent of AS-Na improved the performance of shampoos dramatically.

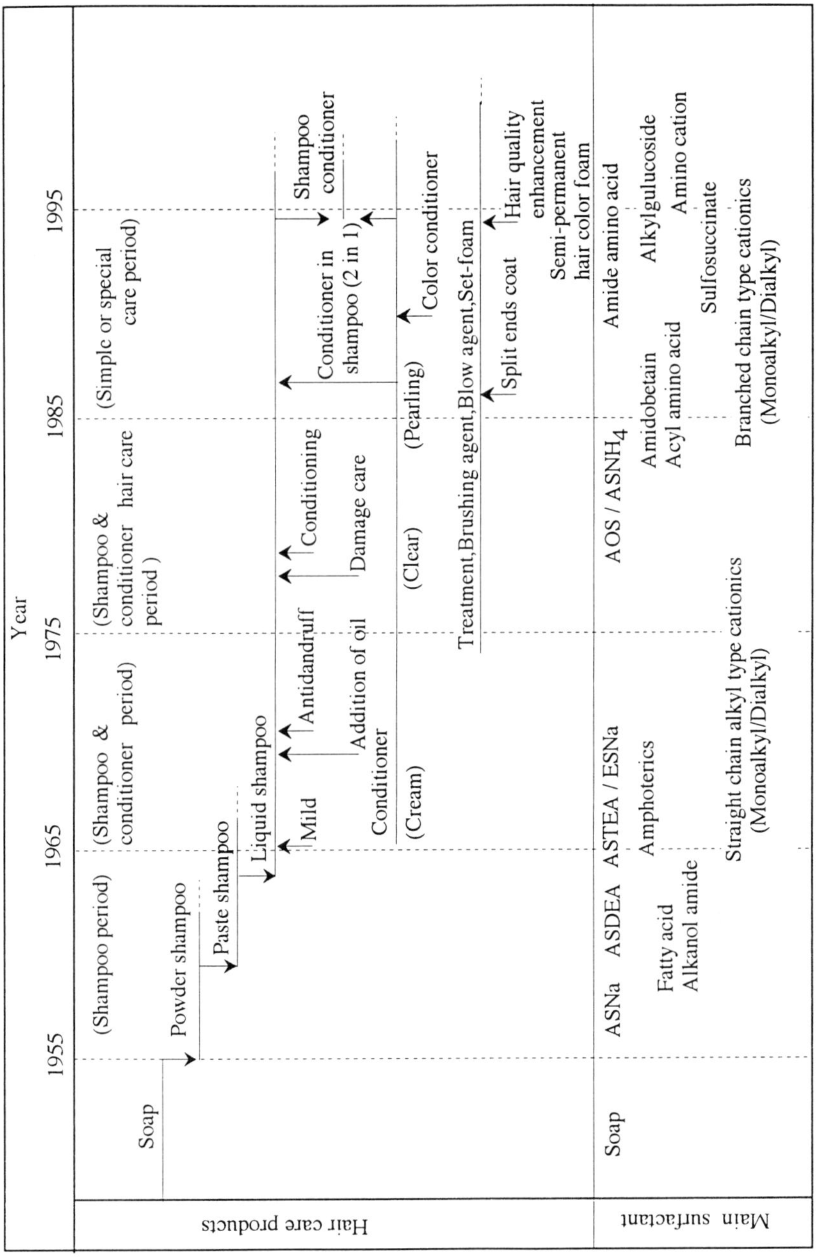

Figure 5.1 History of shampoos and conditioners in Japan.

The AS-Na shampoo was followed by a shampoo in a paste form containing the diethanolamine salt of alkylsulfate (AS-DEA), the detergency of which was lower than that of AS-Na, and by shampoos in a liquid form containing the triethanolamine salt of alkylsulfate (AS-TEA) and the sodium salt of polyoxyethylene alkyl ether sulfate (ES-Na). These last two were much easier to use and formed the basis from which today's sophisticated shampoos were developed.

People started to use conditioner around the mid 1960s; the era of shampoo and conditioner had come. Moreover, shampoo itself was augmented by adding new functions to it so that it would give hair a pleasant feel; oil shampoo and cream shampoo made their way into the market.

Research on the properties of hair itself advanced rapidly from around 1975; microscopic studies on hair, mainly investigating how hair damage occurs physically and chemically, have been painstakingly conducted. The information gained from such studies enabled the development of conditioning shampoos, which minimized the occurrence of damage, deep conditioners, brushing products and blow-dry styling products; hair care products had reached their prime.

The recent market climate is characterized by 'simple to use' and 'specialty function' products, for example hair-coating products that prevent split ends, leave-on or leave-in conditioners, conditioner-in shampoos, colorant-in conditioners, leave-on hair colorants, hair-quality enhancing agents and shampoo-in conditioners. The success of these products is due to exhaustive and extensive research of the functions of surfactants and auxiliary ingredients, and the effects of their complexation. There are also other factors that are thought to boost the use of these products and hence their development:

- an increase in the number of people having long hair
- the increase in hair washing frequency
- the diversity of hair-styles
- a tendency to worry about chemical treatment, such as perming and hair dyeing
- a high awareness of hair damage
- a preference for mild treatment
- a preference for simplicity in use.

Demands from consumers and their values are rapidly changing and researchers must be aware of even subtle changes.

The standard composition of an ordinary shampoo and the formulating purpose of each ingredient are shown in Table 5.2. The fundamental functions required of a shampoo are providing abundant foam, adequate detergency and a comfortable feel for the hair. Low irritancy is also always important. Most shampoos are made up mainly of anionic surfactants with auxiliary, foam-boosting betaines and/or non-ionic surfactants.

Table 5.3 gives the structures of the basic surfactants used in shampoos and their features. Of those listed, polyoxyethylene lauryl ether sulfate is used in the

Table 5.2 General ingredients for shampoos

Ingredient	Function	Standard composition range (wt%)	Example
Surfactant	Cleansing, foaming	10–20	Alkyl sulfate, alkyl ether sulfate, α-olefin sulfonate, amphoteric surfactant
Auxiliary surfactant (foam booster)	Enhance richness of lather, detergency	1–5	Acyl amino acid, alkyl sulfosuccinate, acyl alkanolamide, amine oxide, betaine, soap
Conditioning agent	Feeling of hair, combability	0.1–1	Cationic polymer, silicone, fatty alcohol, protein, cationic surfactant
Thickening agent	Viscosity control	~2	Carboxy vinyl polymer, cellulose derivative
Antidandruff agent, antimicrobial agent	Antidandruff, deodorant	0.5–1.5	ZPT, octopirox, triclosan, sulfur
Opacifier/pearlising agent	Appearance control	~5	Fatty acid glycol ester, latex, metallic soap
Hydrotrope	Solubilization	~10	Propylene glycol, ethyl alcohol, glycerol
Chelating agent	Applicable in hard water	~2	EDTA, citric acid
Preservative	Preservation	~1	Sodium benzoate, paraben
Antioxidant	Stabilizer	q.s.	BHT, α-tocophenol
pH control agent	pH control	q.s.	Phosphoric acid, citric acid
Fragrance	Perfume	q.s.	
Color	Appearance control	q.s.	

q.s. quantum sufficit

Table 5.3 Surfactants used in shampoos

Surfactant	Feature
Alkyl sulfate (AS) $C_{12}H_{25}OSO_3M$ (M = Na, TEA, NH_4)	High foaming Creamy foam
Polyoxyethylene alkyl ether sulfate (ES) $C_{12}H_{25}O(CH_2CH_2O)_nSO_3M$ (n = 2–3; M = Na, TEA, NH_4)	Low foaming when oil is present Light and rough foam
α-Olefin sulfonate (AOS) $RCH{=}CHCH_2SO_3Na$ $RCH(OH)CH_2CH_2SO_3Na$	Light and rough foam
Monoalkyl sulfosuccinate (SS) 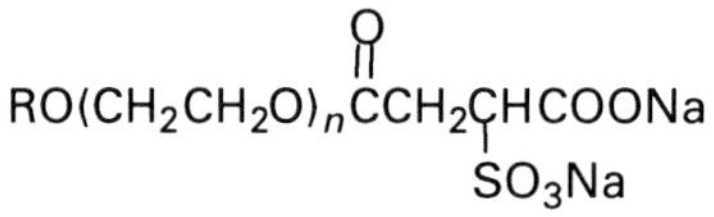 (n = 2–7)	Low irritation Light and creamy foam Low detergency
Acyl amino acid 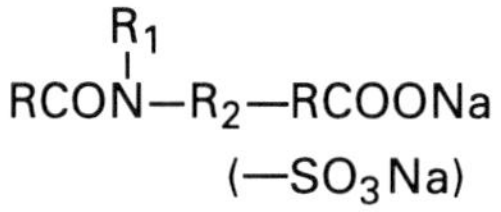 R_1 = H, CH_3; R_2 = CH_2, CH_2CH_2, $CH(CH_2CH_2COOH)$	Low foaming and detergency Low irritation
Betaine alkyl betaine 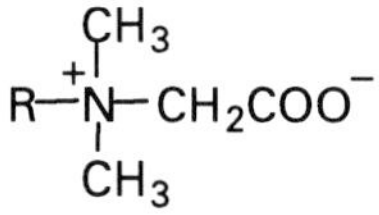 alkylamidopropyl betaine 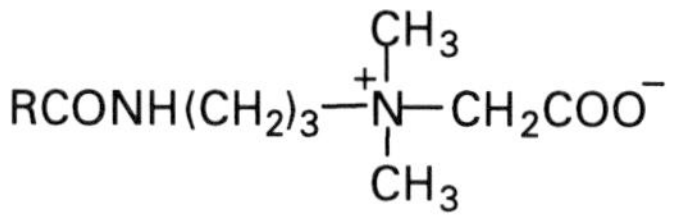	Effective in scum dispersion Low irritation Low foaming and detergency Foam booster
Alkyl imidazoline (amphoteric) 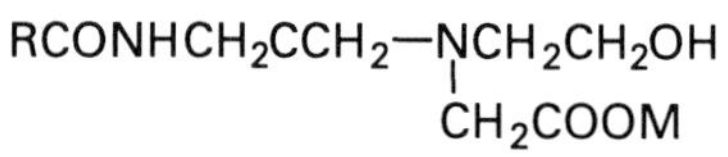 (M = Na, TEA)	Applicable in hard water Low irritation Low foaming and detergency

most formulations. The addition of ethylene oxides (EOs) to lauryl sulfate lowers the Krafft point, giving better stability at low temperatures, and reduces the irritancy, hence the popularity of polyoxyethylene lauryl ether sulfate. Note, however, that foaming power and detergency become weaker the more hydrophilic EO groups are added to lauryl sulfate; usually two to four EOs are added. The effects of EO addition to polyethylene lauryl sulfate on the properties of the sulfate in aqueous solution are summarized in Table 5.4 [8].

Table 5.4 Solution properties of sodium lauryl or laurylpolyoxyethylene sulfates

$C_{12}O(EO)_nSO_3Na$	CMC (M)	Krafft point	Foam volume (ml)[a]
$n=0$	8.6×10^{-3}	16	200
1	4.8×10^{-3}	11	145
2	3.0×10^{-3}	<0	40
3	2.0×10^{-3}	<0	15
4	1.3×10^{-3}	<0	—

[a] Reversed stirred method.

Monoalkyl sulfosuccinates, acyl glutamates, acyl methyl taurates and alkyl imidazolines are low-irritant surfactants. However, they do not possess sufficient foaming power and detergency, and therefore they are used in combination with other surfactants.

The functional properties of surfactants that are most often used are solubilization, emulsification and dispersion, penetration and wetting, detergency, foaming, adsorption and thickening (Table 5.5). Generally, it is rare to incorporate only one kind of surfactant; a combination of surfactants, or surfactants and other ingredients, is used in order to attain target properties. Synergistic effects must also be taken into account. Ways of improving the foaming power of shampoo can be considered as an example of how synergistic effects can improve the function of a shampoo.

Foaming power is an important and basic feature of a shampoo and must be given great attention when formulating a shampoo. Almost any shampoo can foam up well on a dirt-free surface; however, the foaming power decreases dramatically when the shampoo is applied to hair because it has to work on scalp sebum, any hair-care agent already in the hair and/or dust. Consequently, consumers tend to use more shampoo or wash their hair more than once. Auxiliary ingredients are usually incorporated in a shampoo in order to boost the foaming power of the main surfactant. Amphoteric surfactants and non-ionic surfactants such as fatty acid alkanolamides and amine oxides are typical auxiliary surfactants. They form molecular complexes with the main surfactant and synergistically improve foaming power. The stability of foam film (liquid film) should be given much consideration when trying to improve foaming power; stability, and the resulting improvement in foaming power, may be achieved through resistance to the thinning of foam film generated by the static repulsion of ionic groups of surfactants situated on both sides of the film or through the improvement of the compaction of surfactants layered on the film.

Table 5.5 Functions of surfactants and their application in hair-care products

Function	Application
Solubilization	Solubilization of oily ingredients (fragrance, conditioning agent, etc.)
Emulsification/dispersion	Emulsification and stabilization of oil, cleansing
	Dispersion and stabilization of pearlising agent and antidandruff agent (ZPT)
Penetration/wetting	To help active ingredients spread over hair and scalp to exert the effect
Detergency	Cleansing of sebum and soil on hair
Foaming	Building up of foam helps cleansing and prevents redeposition of dirt
	Protection of hair from damage
Adsorption	Adsorbtion onto hair surface, changing its surface properties
	Improvement of combability of hair
Thickening	Increases viscosity by forming structure, easier handling

Hair conditioning is another function currently popular with consumers. Conditioning is achieved by adding a cationic polymer to the shampoo formulation. Of all the cationic polymers used for conditioning, cationic cellulose is the most common. It is assumed that anionic surfactant micelles and cationic cellulose molecules are solubilized by each other in a highly concentrated solution but when they are diluted with an excess of rinsing water phase separation and precipitation of complexes of surfactants and polymers occur. It is thought that the adsorbed precipitate conditions the hair.

The formation of surfactant–polymer complexes and the resulting conditioning effect vary with the type of anionic surfactant and cationic polymer used and the composition. The study of the formation of anionic surfactant–cationic polymer complexes in water has been carried out using phase diagrams [9]. Today, the formulation of a shampoo is aimed at developing a mild shampoo with as strong a conditioning effect as possible. It is important to find the optimum combination of ingredients that can be used for a target functionality; the choice of ingredients greatly affects the properties of a shampoo.

5.4 Conditioners

In Japan people started using conditioner at home when a conditioner containing a cationic surfactant with two long-chain alkyl groups was introduced to the market in 1965. This product slowly but gradually penetrated the market. The instructions for use were typically 'Mix a cap (about 13 ml) of the conditioner well into a quarter of a bowl (about 500 ml) of lukewarm water. Apply to the hair. Rinse off lightly'. Before conditioners with a cationic surfactant became popular, however, people often applied vinegar, lemon juice or another organic acid to the hair in order to rinse off scum caused by soap and calcium in water, and in order to neutralize hair made alkaline by soap. After shampoos containing long-chain anionic surfactants were introduced conditioners containing cationic surfactants started gaining popularity.

While there are many surfactants available for use in shampoos, there are not as many surfactants available for use in conditioners. Researchers have developed a lot of anionic, amphoteric and non-ionic surfactants specifically for shampoos, but as far as surfactants for conditioners are concerned, most are cationic and there is only a handful of different types of cationic surfactants available [10, 11]. This is due partly to the limitation caused by the difficult synthetic process of adding a cationic group to the main structure of the surfactant and partly to the fact that only an amine salt or quaternary ammonium salt can be used as the cationic head of a surfactant.

Conditioner can be rinsed off with water, thus it must be strongly absorbed by the hair in order to function adequately. Conditioners are developed with a focus on their adsorptive power and their ability to alter the quality of hair. The basic functions of a hair conditioner include the prevention of damage on drying hair after washing, minimizing the entanglement of hair, the prevention of moisture loss from dry hair, the reduction of static electricity on hair caused by brushing and the improvement of the feel of hair, e.g. giving it a moisturized and silky feel.

Table 5.6 shows the standard composition of the major ingredients used in conditioners. As can be seen from the table, cationic surfactants represent a sizable portion of a conditioner and most of the conditioner's basic functions are attributable to the characteristics of the cationic surfactants adsorbed onto hair. The cationic surfactants most commonly used in conditioners are quaternary ammonium salts with one or two alkyl chains of 16 to 22 carbons (Table 5.7). Generally, the alkyl groups are in the form of a straight chain and have an even number of carbons because they are industrially derived from fatty acids in natural fats, such as tallow, coconut oil and rapeseed oil. However, recent technology has produced high quality catalysts that can facilitate reactions to derive a long-chain alkyl amine directly from a synthetic alcohol with a branched structure or with an odd number of carbons on the alkyl group [12]. From the long-chain alkyl amine an unconventional cationic surfactant can be derived; this use of catalysts has widened the variety of alkyl groups that can be added to a cationic surfactant. It has been reported that a cationic surfactant with a branched alkyl group of 28 carbons, derived from a Guerbet alcohol with a long alkyl branch, shows interesting properties [13].

Table 5.6 General ingredients for hair conditioners

Ingredient	Function	Standard composition range (wt%)	Example
Cationic surfactant	Softening of hair, improvement of combability	0.3–5	Alkyl trimethyl ammonium chloride Dialkyl dimethyl ammonium chloride
	Antistatic effect	0.5–10	Fatty alcohols, fatty esters, paraffin, silicone, protein
Emulsifier	Emulsification	~2	Polyoxyethylene alkyl ether, monoglycerol
Thickening agent	Viscosity control	~1	Cellulose derivative, polyvinyl pyrrolidone
Antidandruff, antimicrobial agent	Antidandruff/deodorant	0–1	ZPT, octpirox, triclosan, sulfur
Hydrotrope	Solubilization	3–10	Propylene glycol, ethyl alchol, glycerol
Chelating agent	Stabilization	~1	EDTA, citric acid
Preservative	Preservation	~1	Sodium benzoate, paraben
Antioxidant	Stabilization	q.s.	BHT, α-tocopherol
pH control agent	pH control	q.s.	Phosphoric acid, citric acid
Fragrance	Perfume	q.s.	
Color	Appearance control	q.s.	

q.s. = quantum sufficit

Table 5.7 Cationic surfactants used in hair conditioners

Name	Structure	Feature
Alkyl trimethyl ammonium chloride	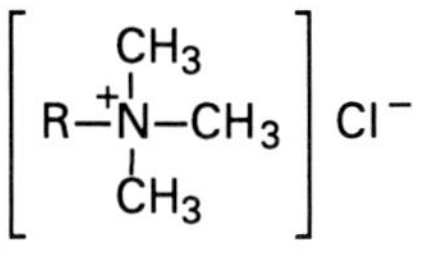 $R = C_{16-18}, C_{20-22}$	Gel formation with higher alcohol Smoothness ($C_{16-18} < C_{20-22}$)
Dialkyl dimethyl ammonium chloride	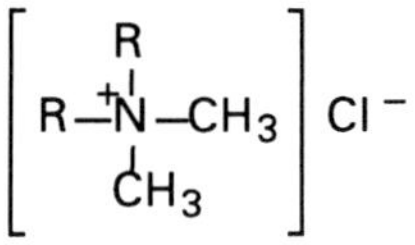 $R = C_{16-18}$	Oily feel ($C_{16-18} > C_{20-22}$) More oily feel than monoalkyl type
Alkyl dimethyl benzyl ammonium chloride	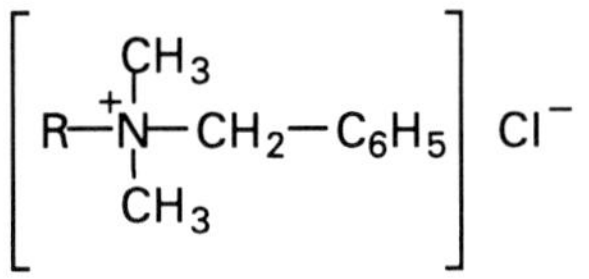 $R = C_{16-18}$	Disinfectant

The Krafft point or gel–liquid crystalline phase transition temperature of a surfactant with a straight chain alkyl group generally becomes higher and the solubility goes down as the alkyl group is extended; surfactants with a long straight alkyl chain are dispersed as hydrated solids (or in a gel phase) at normal ambient temperatures. Thus, they may not be able to develop their capabilities fully at these temperatures. Changing the straight chains to branched chains can drastically reduce the tendency to crystallize; hence, the change can lower the Krafft points or gel–liquid crystalline phase transition temperatures. The following paragraph describes a study that compared anionic surfactants with straight- and branched-chain alkyl groups in terms of solubility in water, adsorption to hair and effect on the physical properties of the hair surface [13]. Table 5.8 lists the structures of the surfactants studied and their abbreviations.

Table 5.8 Molecular structures and abbreviations of cationic surfactants

Surfactant name	Molecular structure	Abbreviation
2-Hexyldecyl trimethyl ammonium chloride	$C_8H_{17}CH(C_6H_{13})CH_2N(CH_3)_3Cl$	g-C_{16}TAC
2-Octyldodecyl trimethyl ammonium chloride	$C_{10}H_{21}CH(C_8H_{17})CH_2N(CH_3)_3Cl$	g-C_{20}TAC
2-Decyltetradecyl trimethyl ammonium chloride	$C_{12}H_{25}CH(C_{10}H_{21})CH_2N(CH_3)_3Cl$	g-C_{24}TAC
2-Dodecylhexadecyl trimethyl ammonium chloride	$C_{14}H_{29}CH(C_{12}H_{25})CH_2N(CH_3)_3Cl$	g-C_{28}TAC
Stearyl trimethyl ammonium chloride	n-$C_{18}H_{37}N(CH_3)_3Cl$	n-C_{18}TAC
Behenyl trimethyl ammonium chloride	n-$C_{22}H_{45}N(CH_3)_3Cl$	n-C_{22}TAC
Distearyl dimethyl ammonium chloride	$(n$-$C_{18}H_{37})_2N(CH_3)_2Cl$	2n-C_{18}DAC

The study of distearyl dimethyl ammonium chloride (2n-C_{18}DAC) and a branched alkyl trimethyl ammonium chloride (g-C_{28}TAC) derived from Guerbet alcohol, was conducted using phase diagrams (Figures 5.2 and 5.3). Liquid crystals started to form at 0°C in the g-C_{28}TAC water system and stayed in that state over a wide range of surfactant concentrations and temperatures. Because of the formation of liquid crystals hair feels slick when conditioner containing these surfactants is applied and when it is rinsed. The formation of liquid crystals stabilizes fat and oil additives in a suspended phase in the conditioner formulation. Table 5.9 lists the gel–liquid crystalline (micellar solution) transition temperatures (T_c) of other cationic surfactants and the appearance of 2.0 wt% solutions at 25°C. Figure 5.4 illustrates the adsorption of cationic surfactants by human hair. With all surfactants the degree of adsorption increases with temperature; in the straight-chain cationic surfactant systems the degree of adsorption starts to increase markedly at the gel–liquid crystalline (micellar solution) transition temperatures (T_c). It is important that the cationic surfactant is in the liquid crystalline state for efficient adsorption by hair; about twice as much of a Guerbet surfactant, which forms liquid crystals at low temperatures, is adsorbed onto hair below 40°C. Figure 5.5 shows the critical surface tension of hair (γ_c)

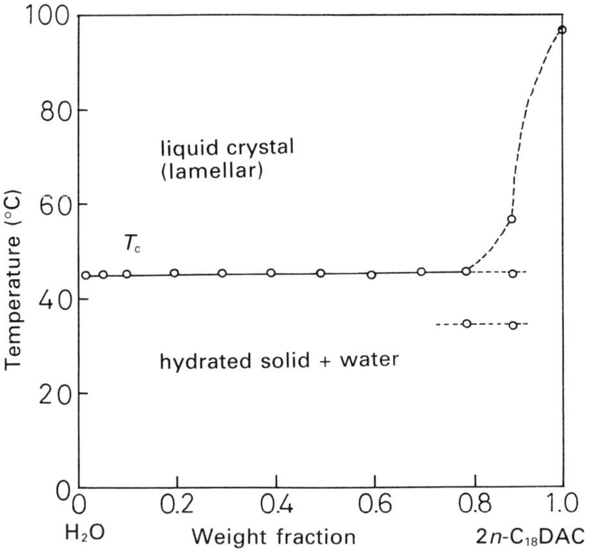

Figure 5.2 Phase diagram of the $2n$-C_{18}DAC–water system. T_c represents the transition temperature between hydrated solid (gel) and liquid crystalline phases.

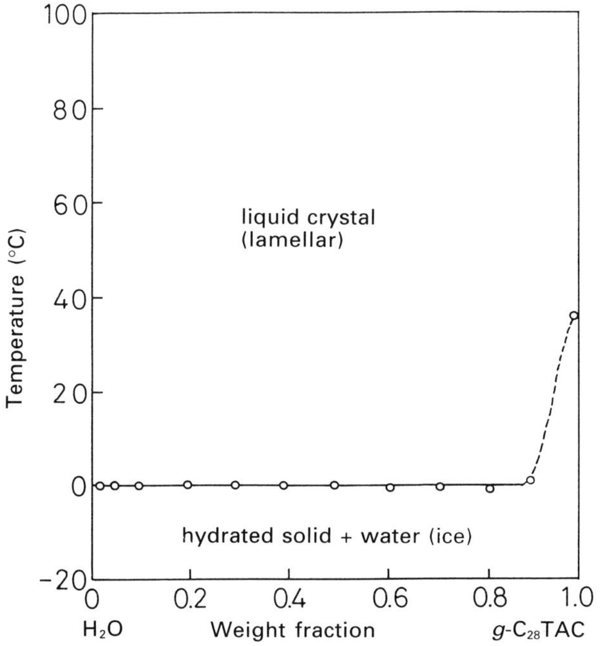

Figure 5.3 Phase diagram of the g-C_{28}TAC–water system.

treated with cationic surfactant solutions. The value of γ_c decreases as the adsorption of the cationic surfactant increases; the adsorption increases the hydrophobicity of the surface of hair. The kinetic frictional coefficient (μ_k) of hair fibers treated with cationic surfactant is shown in Figure 5.6. The μ_k values of wet and dry hair correlate with the γ_k value (Figure 5.7); as the surface energy decreases, the μ_k values decrease and the smoothness of the hair increases. Among the surfactants studied the Guerbet cationic surfactant with an alkyl group of 28 carbons works best in terms of smoothness.

Table 5.9 Solution properties of various cationic surfactants in water

Surfactant	CMC (M)	T_c (°C)	Appearance of 2.0 wt% solution at 25°C
g-C_{16}TAC	6.3×10^{-3}	<0	Micellar solution
g-C_{20}TAC	3.5×10^{-4}	<0	Liquid crystal (lamellar)
g-C_{24}TAC	3.2×10^{-5}	<0	Liquid crystal (lamellar)
g-C_{28}TAC	3.3×10^{-6}	<0	Liquid crystal (lamellar)
n-C_{18}TAC	2.4×10^{-4}	18	Micellar solution
n-C_{22}TAC	2.5×10^{-5}	47	Hydrated solid (gel)
2*n*-C_{18}DAC	2.5×10^{-6}	45	Hydrated solid (gel)

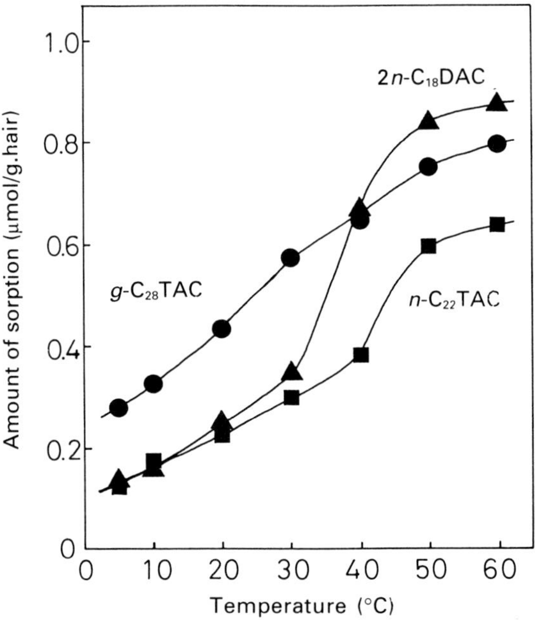

Figure 5.4 The sorption of cationic surfactants by normal hair as a function of temperature. Sample solution concentration was kept constant at 2.0 wt % and the immersion time was constant at 10 min.

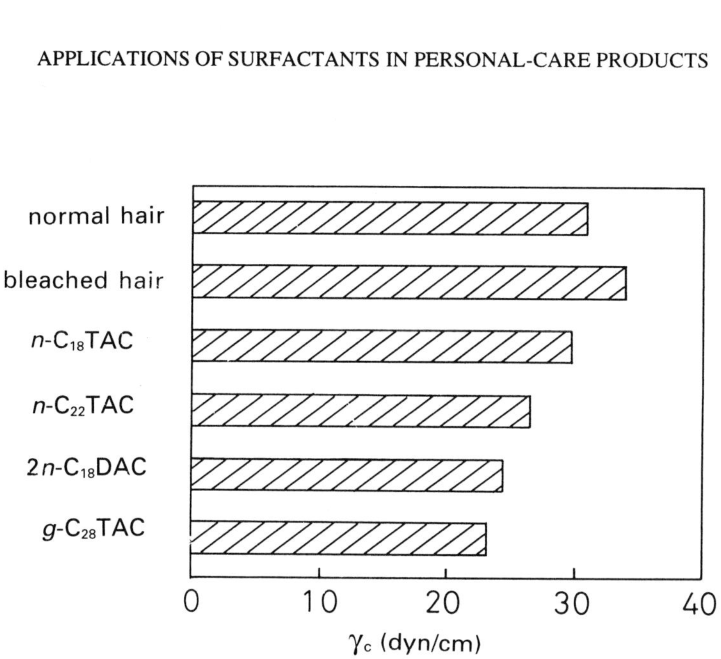

Figure 5.5 The critical surface tension (γ_c) of hair treated with cationic surfactants.

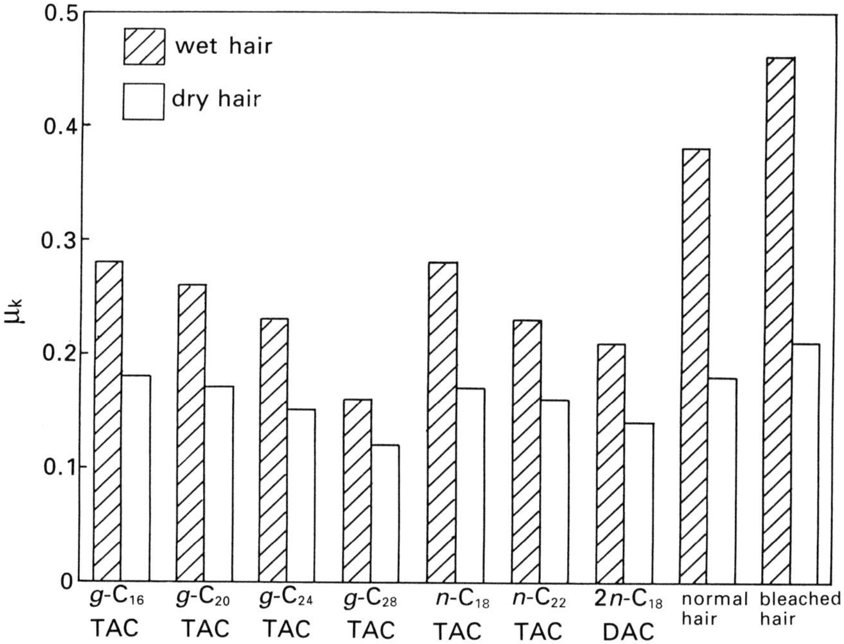

Figure 5.6 The kinetic frictional coefficient (μ_k) of hair treated with cationic surfactants.

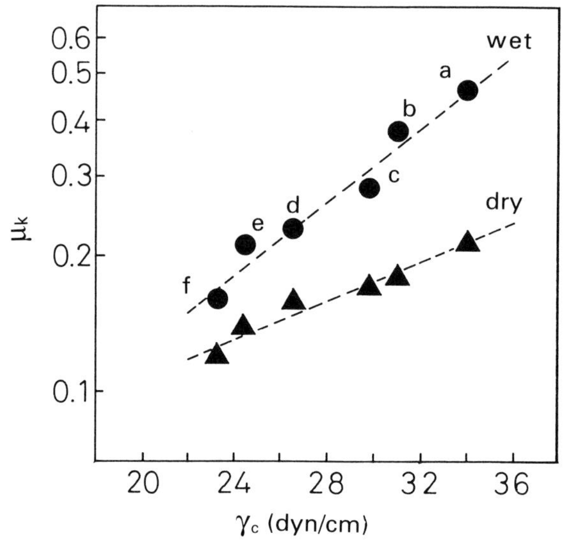

Figure 5.7 The relationship between the kinetic frictional coefficient (μ_k) and critical surface tension (γ_i) for wet (●) and dry (▲) hair: (a) bleached hair; (b) normal hair; (c) n-C_{18}TAC; (d) n-C_{22}TAC; (e) $2n$-C_{18}DAC; (f) g-C_{28}TAC.

There has been a large number of reports examining the adsorption of commonly used cationic surfactants by hair [14–18]. The following is a summary of some of these reports:

1. Adsorption (strictly 'sorption') increases with the immersion time of hair in a cationic surfactant solution. Sorption increases rapidly for the first ten minutes and then continues to increase gradually. It is thought that adsorption reaches saturation point quickly and that slow permeation of cut sections and openings between cuticles follows.
2. In low concentrations, up to 1%, the sorption of cationic surfactant increases with concentration.
3. The sorption of cationic surfactant increases as the immersion temperature increases. It is postulated that the phenomenon is not just a matter of a change in ionic interaction between the hair surface and the cationic surfactant but also involves a structural and permeability change in hair.
4. The sorption of cationic surfactant increases with the pH of the solution. This is because the negatively charged sites of protein in hair increase as the pH value of the solution goes up. Reverse pH dependence is observed with anionic surfactants.

5. In a comparison of cationic surfactants of the same skeleton with alkyl groups of 8 to 18 carbons, a longer surfactant is observed to have more sorption.
6. Chemically damaged hair, such as permed or bleached hair, sorbs more cationic surfactant than normal hair.
7. Hair tends to be smoother and more antistatic the more cationic surfactant is sorbed onto hair.

New types of surfactants have been developed. An asymmetrical dialkyl quaternary ammonium salt with a tallow alkyl chain and a Guerbet branched octyl chain is now available for use in conditioners; the branching makes the dialkyl cationic surfactant more soluble [19]. Tricetyl methyl ammonium chloride [20], insoluble in water, has better sorption than a conventional cationic surfactant; it gives a good conditioning effect and a comb runs well through hair with the surfactant sorbed onto it. The synthesis of quaternary ammonium salts of long-chain triglycerides, without breaking the triglyceride skeleton, has been successful and it has been reported that formulations with these synthetic salts make good skin-care lotions and conditioners that improve the smoothness of hair [21]. A methyl glucoside with 10 moles of ethoxylate linked to hydroxypropyl dimethyldodecyl ammonium chloride has been developed [22]. This formulation has low eye irritation and is ideal as hair and skin conditioner. Furthermore, a cationic surfactant with a guanidino cationic group has been developed [23]. It is inferred that the surfactant remains on hair tenaciously because the guanidino group has a steric structure that makes it easy for it to be attracted to an anionic, carboxyl group of hair and it also has a secondary amide group that can easily form a hydrogen bond with the carboxyl groups. It has been reported that a conditioner incorporating this surfactant gives hair characteristic smoothness during rinsing and adequate moisture after the hair dries [23].

As is mentioned above, cationic surfactants are designed to be adsorbed onto hair strongly and to give a good conditioning effect. Some of the conditioners on the market are targeted to the reduction of irritancy; they are designed with the intention of providing a mild shampoo. In recent years, a totally different kind of approach has been taken in the development of cationic surfactants; a cationic surfactant with a high foaming power has been developed and utilized in a conditioner formulation, marketed as a shampoo-in conditioner [24].

Fats and oils, as well as cationic surfactants, are important ingredients for conditioners. These additives are adsorbed onto hair along with cationic surfactants, giving smoothness and a moisturized feel to hair. Long-chain alcohols with 16 to 18 carbons give a slick oily feel during the application of conditioner, rinsing and towel drying. They form a gel structure with cationic surfactants [25], giving the characteristic smooth oily feel. Silicone derivatives, particularly water-insoluble, highly polymerized dimethyl polysiloxanes, are also receiving attention. Hydrophobic silicone derivatives are adsorbed onto hair by virtue of hydrophobic forces, but their behavior is rather different from that of hydrocarbons.

Damaged hair, such as permed or bleached hair or the tip of a long hair, has high energy with high hydrophilicity; such hair does not adsorb and retain as much hydrophobic oil as normal hair [26–28]. In one report, an increase in the adsorption of isopropylmyristate (IPM) was observed as hair became damaged [27]; cationic surfactants tend to make the damaged part more hydrophobic by adsorption. A decrease in absorption of dimethyl polysiloxane was also observed in damaged hair. Both of these observations relate to increased hydrophilicity and give useful information for developing a better formulation.

5.5 Skin-care products

Skin-care products, embracing body cream and body milk as the most general types, remain in contact with skin for a long time therefore only a limited number of low irritant surfactants can be used. Extensive research has been carried out to develop even milder surfactants.

Skin-care products are becoming more and more functional and among the most sought-after functions are high humectant ability and the retention of various effects. While the search for ingredients with such functions goes on, more cream and milk products are formulated to meet the demands of each age group and the needs of each season; their compositions can be freely altered in an oil-in-water (o/w) or water-in-oil (w/o) emulsifying form. By incorporating a variety of humectants and oils, manufacturers are able to develop and sell products that help skin to recover its basic function of retaining sufficient water and as a result to keep its supple condition.

Recently, formulation has been carried out in a way that enables researchers to optimize the composition of a product while focusing on discovering an unknown property of an existing active ingredient rather than finding a new ingredient. The aim of this is to try to make these active ingredients work more efficiently on the skin.

It has been found that intercellular lipid lamellae, predominantly ceramide, in stratum corneum located on the surface of the skin play an important role in maintaining healthy skin [29–31]. Along with cell-membrane-constructing phospholipids and intracellular ceramides, the intercellular ceramides exemplify the important contribution that naturally occurring lipids make to living organisms. Synthetic ceramides that are similar to the natural intercellular ceramides have been developed and, reportedly, a product incorporated with some of these ceramides enables damaged skin to recover to its normal state, as the natural ceramides do, by replenishing intercellular space with the pseudo-ceramide and reconstructing lost lamellae [32].

Artificial intercellular lipids with synthetic pseudo-ceramides remain in a paste form up to very high temperatures therefore, in order to stabilize them in a suspended form, the lipids are formed into fine multi-lamellar droplets by a liquid crystal emulsion method [33]. The lipids stabilized in this emulsion form have been known to possess a higher water-holding ability than in a normal emulsion form [34]; it is important that the product form is optimized for efficiency.

Products containing a quasi-liposome form of cell-membrane-constructing phospholipid, another lamellar lipid, have been developed. The form taken by the liposome has been proposed as an efficient drug delivery carrier and extensive medical research on this has been carried out, with special reference to percutaneous absorption and drug release.

Ordinarily, active ingredients are dissolved in one of the two phases of an o/w or w/o system when developing a milk or cream type of cosmetic product. In addition, a new type of cosmetic form using lamellar liquid crystals has been reported [35]. These new cosmetics are totally different from conventional emulsified cosmetics. They take the form of a lamellar liquid crystalline gel and the gel itself is an active ingredient. The strata formed by the gel, in the same configuration as stratum corneum intercellular lipids, are capable of covering the surface of skin to reduce the transpiration of water retained in the skin and to continuously refill stratum corneum with their own water, which is held rigidly by the lamellae. These cosmetics are unique in that these effects are not expected from an o/w emulsion, which allows transpiration from skin, and a w/o emulsion, where water is not in direct contact with skin. Furthermore, these cosmetics transport the active ingredient into stratum corneum quickly because they are formulated in a similar way to the stratum corneum intercellular lipids.

As well as these new cosmetics, it has been reported that cosmetics with saturated fatty acid monoacyl glycerine in lamellar crystals increase the water content of stratum corneum [36]. With these advances in cosmetic formulation, it is assumed that there will be many more new skin-care products in lamellar crystal form, totally different from the conventional emulsified ones.

In summary, recent research on the application of surfactants to skin-care products is oriented towards the study of three-dimensional structures, such as liposome and lamellar liquid crystals, as well as conventional areas, such as emulsifying properties. It is likely that rather than designing new surfactants the industry will continue to search for unknown attributes of existing surfactants that can be applied to new cosmetic products.

5.6 Conclusion

This chapter has discussed the development and application of surfactants, together with the related historical background of Japan. It is presumed that demands for multi-functional cosmetic products will increase as people's values and environments change. Researchers should be prepared in advance so that they can quickly respond to the changing demands of the market. Additionally, they must pay attention to fundamental nature-friendly principles as the need to ensure the safety of aquatic living organisms, the development of surfactants that are easily broken down by microorganisms, using materials that can be recycled and devising ways in which quality materials can be manufactured with low energy input and at a low cost. The evolution of personal-care products depends heavily on how researchers in this line of work manage to allow nature and people's luxury to coexist in harmony.

References

[1] Yahagi, K. and Iwai, H. (1996) Application of surfactants in personal care products. *Journal of the Japanese Oil Chemists' Society*, **45** (10) 1133-43.

[2] Kurosaki, T., Imokawa, G. and Ishida, A. (1987) The development of monoalkyl phosphate as a low skin irritating anionic surfactant. *Journal of the Japanese Oil Chemists' Society*, **36** (9) 629-37.

[3] Kawai, M. and Imokawa, G. (1984) The induction of skin tightness by surfactants. *Journal of the Society of Cosmetic Chemists*, **35** 147-56.

[4] Imamura, T., Itaya, H. and Nakajima, M. (1995) Shampoo and body shampoo in Japan. *Perfume Japan*, **187** 69-80.

[5] Sakamoto, K. (1995) Surfactants derived from amino acids. *Journal of the Japanese Oil Chemists' Society*, **44** (4) 256-65.

[6] Kanari, M., Kawasaki, Y. and Sakamoto, K. (1993) Acylglutamate as an anti-irritant for a mild detergent system. *Journal of the Society of Cosmetic Chemists Japan*, **27** (3) 498-505.

[7] Sano, K. and Koyama, M. (1995) Developments and applications of new amino acid based surfactants. *Fragrance Journal*, **23** (4) 87-96.

[8] Tsujii, K., Okahashi, K. and Takeuchi, T. (1981) Physicochemical properties of anionic surfactants with poly(oxyalkylene) group in water. *Journal of the Japanese Oil Chemists' Society*, **30** (9) 566-72.

[9] Goddard, E.D. (1990) Polymer/surfactant interaction. *Journal of the Society of Cosmetic Chemists*, **41** 23-49.

[10] Quack, J.M. (1976) Quaternary ammonium compounds in cosmetics. *Cosmetics and Toiletries*, **91** (2) 35-52.

[11] Gerstein, T. (1979) *Cosmetics and Toiletries*, **94** (11) 32-41.

[12] Okabe, K. and Abe, H. (1988) Recent trends of catalysts for production of aliphatic tertiary amines. *Journal of the Japanese Oil Chemists' Society*, **37** (7) 485-91.

[13] Yahagi, K., Hoshino, N. and Hirota, H. (1991) Solution behaviour of new cationic surfactants derived from Guerbet alcohols and their use in hair conditioners. *International Journal of Cosmetic Science*, **13** 221-34.

[14] Scott, G.V., Robbins, C.R. and Barnhurst, J. (1969) Sorption of quaternary ammonium surfactants by human hair. *Journal of the Society of Cosmetic Chemists*, **20** 135-52.

[15] Finkelstein, P. and Laden, K. (1971) The mechanism of conditioning of hair with alkyl quaternary ammonium compounds. *Applied Polymer Symposium*, **18** 673-80.

[16] Tokiwa, F., Hayashi, S. and Okumura, T. (1976) Hair and surfactants, in *Biology and Disease of the Hair* (eds. K. Toda, Y. Ishibashi, Y. Hori and F. Morikawa), University of Tokyo Press, Tokyo, pp. 631-40.

[17] Harusawa, F., Kato, T. and Tanaka, M. (1981) The sorption of surfactants by human hair and textile fibers and the kinetic frictional coefficient of human hair. *Journal of the Society of Cosmetic Chemists Japan*, **15** (3) 225-32.

[18] Ohbu, K., Tamura, T., Mizushima, N. and Fukuda, M. (1986) Binding characteristics of ionic surfactants with human hair. *Colloid and Polymer Science* **264**, 798-802.

[19] Jurczyk, M.F. (1991) A new quaternary conditioner for damaged hair. *Cosmetics and Toiletries*, **106** (11) 91-95.

[20] Jurczyk, M.F., Berger, D.R. and Damasco, G.R. (1991) Quaternary ammonium salt. *Cosmetics and Toiletries*, **106** (4) 63-68.

[21] Baydar, A. and Johnston, R. (1991) Novel quaternary ammonium salts derived from triglycerides and their application in skin and hair products. *International Journal of Cosmetic Science*, **13** 169-90.

[22] Polovsky, S.B. (1991) An alkoxylated methyl glucoside quaternary. *Cosmetics and Toiletries*, **106** (7) 59-65.

[23] Mitamura, J., Suzuki, N., Onuma, K., Miyake, M., Nakamura, T. and Kiyomiya, A. (1994) Development of a new cationic surfactant for use in high quality hair conditioners, *Proceedings of the 18th IFSCC Congress Venezia Preprint (Poster Presentation)*, International Federation Society of Cosmetic Chemists, Venice, pp. 202-21.

[24] Yahagi, K. (1995) Development and applications of cationic surfactants for hair care products. *Fragrance Journal*, **23** (4) 55-64.

[25] Noda, A., Yamaguchi, M., Machida, Y. and Fukushima, S. (1986) Structure of gels composed of cationic surfactant, long chain alcohol and water. *Journal of the Society of Cosmetic Chemists Japan*, **20** (2) 103-10.

[26] Yahagi, K. (1992) Silicones as conditioning agents in shampoos. *Journal of the Society of Cosmetic Chemists*, **43** 275-84.

[27] Ishida, A. and Watanabe, H. (1977) Emollient ingredients in hair preparations. *Fragrance Journal*, **5** (25) 42-47.

[28] Watanabe, S. and Yahagi, K. (1995) Sorption behavior of silicone on hair. *Journal of the Society of Cosmetic Chemists Japan*, **29** (1) 64-68.

[29] Elias, P.M. (1981) Lipids and the epidermal permeability barrier. *Archives of Dermatological Research*, **270** 95-117.

[30] Lampe, M.A., Burlingame, A.L., Whitney, J., Williams, M.L., Brown, B.E., Roitman, E. and Elias, P.M. (1983) Human stratum corneum lipids: characterization and regional variations. *Journal of Lipid Research*, **24** 120-30.

[31] Elias, P.M. (1983) Epidermal lipids, barrier function, and desquamation. *Journal of Investigative Dermatology*, **80** (6) 44s-49s.

[32] Imokawa, G. (1995) Structure and function of intercellular lipids in the stratum corneum. *Journal of the Japanese Oil Chemists' Society*, **44** (10) 751-66.

[33] Suzuki, T., Tsutsumi, H. and Ishida, A. (1983) Secondary droplets formed in O/W emulsion: formation mechanism and effects of their formation on the properties of emulsion. *Journal of the Chemical Society Japan*, **3** 337-44.

[34] Suzuki, T., Fukasawa, J., Iwai, H., Sugai, I., Yamashita, O. and Kawamata, A. (1993) Multilamellar emulsion of stratum corneum lipid: formation mechanism and its skin care effects. *Journal of the Society of Cosmetic Chemists Japan*, **27** (3) 193-205.

[35] Iwai, H., Fukasawa, J., Fukuda, M., Sugai, I., Uchida, H. and Suzuki, T. (1996) Application of liquid crystal to skin care cosmetics and its properties. *Journal of the Society of Cosmetic Chemists Japan*, **30** (3) 310-20.

[36] Sone, T., Hanamizu, T., Yoshikawa, S., Ichioka, M. and Yokokura, T. (1994) Moisturizing effect of lipids forming lamellar structures. *Journal of the Japanese Cosmetic Science Society*, **18** (4) 186-92.

6 Methyl ester sulfonates

Teruhisa Satsuki

6.1 Introduction

Methyl ester sulfonates (MESs) are oleochemical-based anionic surfactants derived from oil palm or coco through trans-esterification and subsequent sulfonation. Linear alkyl benzene sulfonates (LASs), alkyl sulfates (ASs) and alpha olefin sulfonates (AOSs) are generally used as detergent surfactants. However, since the two oil crises in the 1970s, detergent manufacturers have been interested in natural fat- and oil-based surfactants, rather than those that are petroleum based, and have considered MESs as a possible detergent ingredient.

MESs have good surface-active properties [1, 2] and excellent detergency performance as a laundry detergent main ingredient [3–5], as well as good biodegradability [6, 7]. MESs were shown to be scum dispersant in the 1960s, and thus their sulfonation mechanism [8, 9], characteristics, application and manufacturing process [10, 11] have been extensively studied. They have only recently been found to be applicable as a main component of laundry detergent products and their manufacture on a commercial basis is now possible. There are problems in manufacturing MESs because of (a) the dark color of the sulfonated product, (b) the generation of disodium salts through sulfonation as a by-product and (c) the low active ingredient content for a detergent raw material. It is thus difficult to produce high quality MESs suitable for laundry detergents. Due to hydrolysis MESs may possibly undergo degradation into the disodium salt on contact with alkaline components [12].

The technology became available for producing high quality MESs on a commercial basis in the early 1990s through improvements such as sulfonation, bleaching and neutralization. This new technology was introduced into Japan, and a new compact powder detergent was produced in 1991. Further compact products were subsequently developed using the improved MESs. In the USA Stepan produced MESs suitable for dish-washing detergent products in the 1980s [13, 14].

This chapter summarizes the application of MESs in laundry detergent formulations, their fundamental surfactant properties, their solid structure and their biodegradability. All the research results were obtained from work carried out at the Lion Corporation.

6.2 Basic surfactant properties

MESs have the chemical structure shown in Figure 6.1. The structure is that of a fatty acid methyl ester with a sulfonate group in the α position. MESs are prepared by the direct sulfonation of fatty acid methyl esters from triglyceride through trans-esterification. The number of carbon atoms in MESs is generally 12–18 for detergent use.

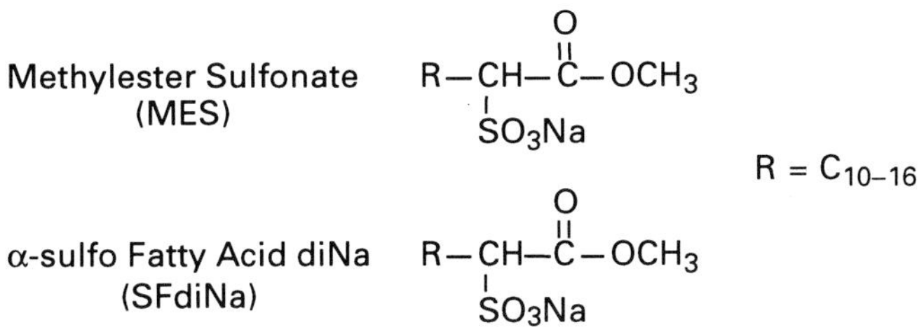

Figure 6.1 Chemical structures of MESs and SFdiNa.

Fujiwara reported the basic properties of MESs and an assessment of their water hardness tolerance [15].

Solubility and CMC curves as a function of temperature for C_{14}-, C_{16}- and C_{18}MES-Na are shown in Figure 6.2 along with those of the corresponding calcium salts. The CMCs and Krafft points of MES-Na, MES-Ca and AS-Na (sodium lauryl sulphate) are presented in Table 6.1. The solubility and CMC were found to decrease with rising carbon number and a counterion change from Na to Ca. These features are characteristic of anionic surfactants [16]. Table 6.1 shows that the increase in Krafft points of the MES compounds as a result of the counterion change is less than that for the AS compounds [16, 17].

Table 6.1 Krafft points and CMC of some surfactants

Surfactant	Krafft point (°C)	CMC (mM)	Reference
C_{14}MES-Na	6	2.8 (13°C)	
C_{16}MES-Na	17	0.73 (23°C)	
C_{18}MES-Na	30	0.18 (33°C)	
C_{14}MES-Ca	28	0.66 (30°C)	
C_{16}MES-Ca	41	0.19 (45°C)	
C_{18}MES- Ca	49	0.042 (50°C)	
C_{12}AS-Na	9	8.1	[17]
C_{14}AS-Na	30	2.1	[17]
C_{12}AS-Ca	50	1.2	[16]
C_{14}AS-Ca	71	0.68	[16]

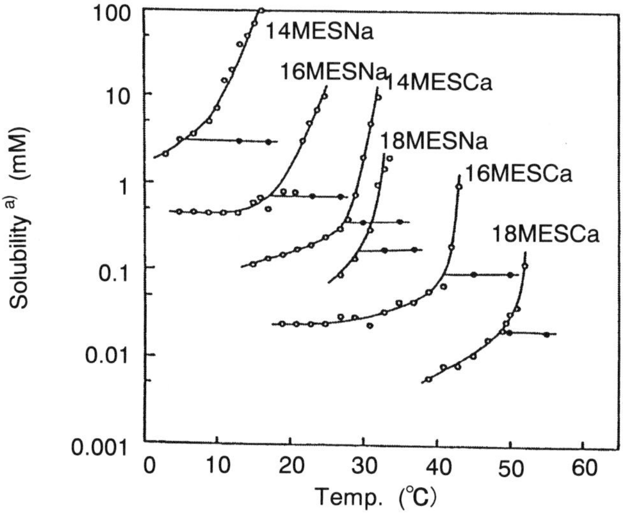

Figure 6.2 Solubility and CMC curves of MES surfactants. Concentrations of MES-Ca were all represented as $RCH(COOCH_3)SO_3.½Ca$.

The CMC of MES-Ca is approximately 25% that of the Na salt. Using the method of Shinoda, the logarithm of CMC was plotted against hydrophobic chain length (Figure 6.3) and the logarithm of counterion concentration [18]. Good linearity for MES-Na and MES-Ca was demonstrated, suggesting micelle formation by MES-Na and MES-Ca following the equation:

$$\ln(\mathrm{CMC}) = -m\omega/kT + K_g/Z_i\{\ln(2000\pi\sigma^2/DNkT) - \ln C_g\} + \mathrm{constant}$$

where m is the number of carbon atoms, ω is the cohesive energy change for transferring one methylene group from a hydrophobic environment to aqueous medium, K_g is an experimental constant related to the degree of counterion binding of micelles, K_g/Z_i is determined from the slope of the CMC as a function of Z_i valent counterion concentration, D is the dielectric constant of the solution, N is Avogadro's number, σ is the charge density on the micelle surface and C_g is the total concentration of the counterions per litre. The energy calculated for MES-Na and MES-Ca was $1.1kT$ and $0.93kT$, respectively, and these values are almost equal to that of a typical ionic surfactant (e.g., AS-Na, $1.1kT$).

The dissociation of counterions for MES-Na and MES-Ca micelles was determined and was found to be essentially the same as that of typical anionic surfactants.

The aggregation number of micelles is important for expressing the physico-chemical properties of surfactant solutions. The micelle weight, aggregation number and second virial coefficient for MES-Na and MES-Ca are listed in

Table 6.2. The aggregation number of C_{14}MES increased from 81 to 119 and the second virial coefficient decreased from 7.30×10^{-3} to 3.05×10^{-4} with increasing electrolyte concentration. This behavior corresponds to that of C_{12}AS-Na reported by Hayashi [19]. The aggregation number of C_{14}MES is basically the same as that of C_{16}MES, whereas C_{18}MES number slightly exceeds that of C_{16} in the 0.01 N $NaNO_3$ solution. The aggregation number of MES-Ca is larger than that of MES-Na, even in distilled water, and increases with electrolyte addition.

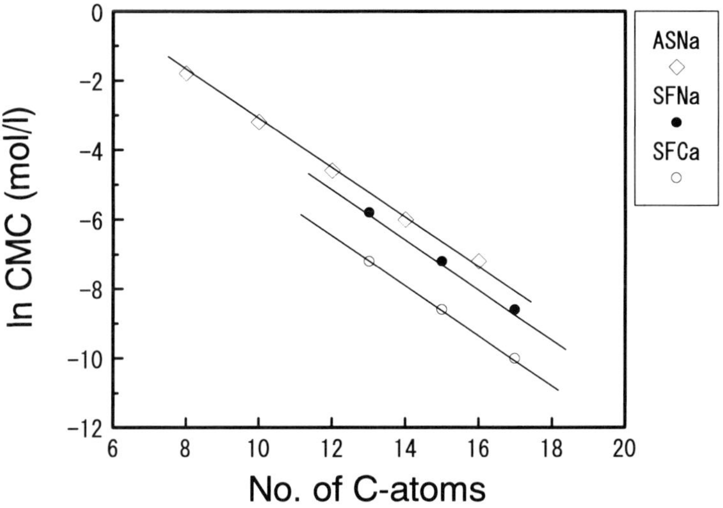

Figure 6.3 Relationship between the logarithm of the CMC and the number of carbon atoms in the hydrophobic chain. The number of carbon atoms was plotted according to the following structures: AS = $C_nH_{2n+1}OSO_3M$; SF = $C_{n-1}H_{2(n-1)+1}CH(COOCH_3)SO_3M$. Key: AsNa ◊, SFNa ●, SFCa O

Table 6.2 Micellar weights, aggregation numbers and second virial coefficients (B_2) of some surfactants.

Surfactant	Medium	Micellar weight	Aggregation number	B_2 (ml/g)
C_{14}MES-Na	0.01 N $NaNO_3$	28 000	81	7.30×10^{-3}
	0.1 N $NaNO_3$	32 800	95	8.73×10^{-3}
	0.4 N $NaNO_3$	41 000	119	3.05×10^{-4}
C_{16}MES-Na	0.01 N $NaNO_3$	31 600	85	5.20×10^{-3}
C_{18}MES-Na	0.01 N $NaNO_3$	42 500	106	2.18×10^{-3}
C_{14}MES-Ca	H_2O	32 600	96	1.87×10^{-3}
	0.003 N $Ca(NO_3)_2$	45 100	132	7.78×10^{-4}
	0.01 N $Ca(NO_3)_2$	42 200	124	-2.39×10^{-4}

6.3 Assessment of water hardness tolerance

Fujiwara proposed a mechanism for the good hardness tolerance of MESs based on the precipitation phase boundary for MES Ca salts [15]. The boundaries for C_{12}AS-Na, C_{14}MES-Na and C_{16}MES-Na are shown in Figure 6.4. The boundary lines were observed after 10 min, 1 h, 1 day and in the equilibrium state, since precipitation was time dependent. The equilibrium phase boundary for C_{12}AS-Na, which has poor tolerance to hardness, agreed with that reported previously [20]. The phase boundary at equilibrium was quite close to that at 10 min, indicating that precipitation occurs very rapidly and equilibrium is attained easily. In ordinary washing this surfactant would precipitate within 10 min and surface activity would be lost.

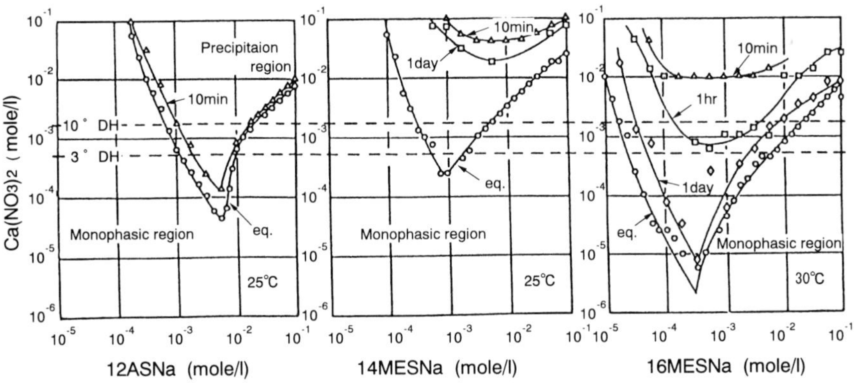

Figure 6.4 Precipitation phase boundary diagrams of C_{12}AS-Na, C_{14}MES-Na and C_{16}MES-Na.

Precipitation region concentrations for C_{14}MES-Na at equilibrium is low and that for C_{16}MES-Na is spread over the range 3–10°DH (DH is German hardness: 1°DH=10 ppm CaO). If washing is under the control of the precipitate boundary in equilibrium, all MES would precipitate with loss of solution activity. The precipitation region of C_{14}MES-Na and C_{16}MES-Na in an actual washing was much smaller than that of AS-Na. The reason for the hardness tolerance of MES-Na may be due to the extremely slow precipitation rate of Ca salts at low temperature.

6.4 Solid state properties

Few studies have been carried out on the solid or crystalline state of anionic surfactants apart from the monocrystalline X-ray diffraction analysis of the hydrated solid of sodium dodecyl sulfate [21]. To extend this to detergent powders, the crystalline state must be understood since features such as powder caking and the quick dissolving rate of a powder particle in water are closely related to the surfactant solid properties.

Fujiwara investigated the crystalline state of MESs with regard to hydration, phase transition and ad/desorption of moisture using the sodium salt of C_{16}MES as a model substance [22].

The phase behavior of MESs was studied by differential scanning colorimetry (DSC) for several MES solid samples with different moisture contents. The results for samples with 0 to 8.9% moisture content are shown in Figure 6.5. For non-hydrated MESs a single endothermic peak was noted. The phase transition temperature (T_c) and enthalpy change (ΔH) were 112°C and 22.1 kJ/mol, respectively. With increasing moisture the endothermic peak gradually shifted to lower temperatures with a concomitant decrease in size. At the same time, a new peak appeared at 68 °C and the size of this peak increased at constant temperature. The spectrum at 8.9% moisture content showed an endothermic peak with T_c and ΔH of 7 °C and 67.5 kJ, respectively, which may possibly have been due to the phase transition of MES.2H_2O crystals from solid to liquid form since the molar ratio of [H_2O]/[MES] was 2.0.

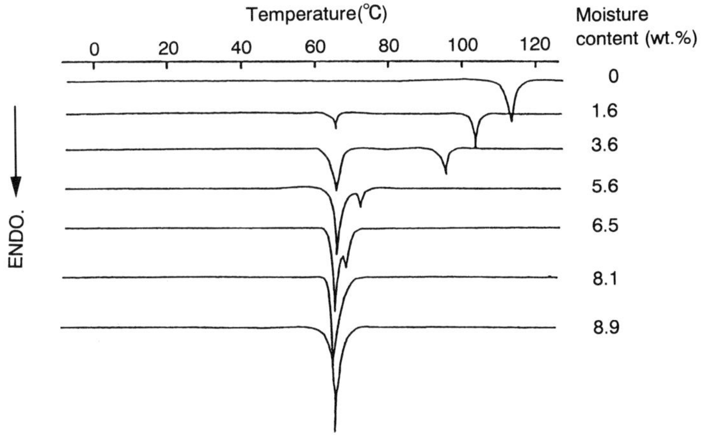

Figure 6.5 DSC curves for MES solids with moisture contents from 0 to 8.9 wt%.

The X-ray diffraction patterns for these MES samples are presented in Figure 6.6. The MES solid at 0 and 8.9% moisture content gave one peak corresponding to spacings of 26.9 and 30.1 Å, respectively. At 1.6 and 5.6% moisture content, however, two peaks of 26.9 and 30.1 Å appeared simultaneously. The two different crystal structures thus appear to coexist independently in solid MESs with moisture contents from 0 to 8.9%.

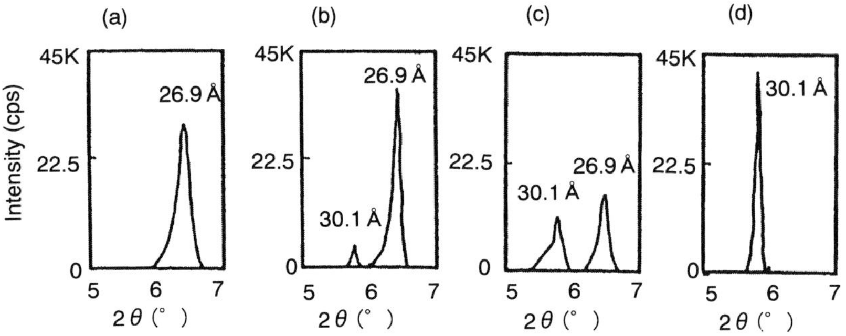

Figure 6.6 X-ray diffraction patterns of MES solids with moisture contents from 0 to 8.9 wt%: (a) 0 wt%, (b) 1.6 wt%, (c) 5.6 wt%, (d) 8.9 wt%.

Fujiwara analyzed the phase transition behavior quantitatively using the phase rule and the Flory–Huggins–Scott equation [23]. MES solids with 0 to 8.9% moisture content were found to be a mixture of MES crystals of anhydrate and 2 mole hydrate.

MES solids with greater moisture contents were also investigated. DSC of MES solids at 8.9 to 32.6% moisture content was carried out and the results are shown in Figure 6.7. Data similar to those in Figure 6.6 were obtained, showing a shift of the endothermic peak corresponding to the phase transition of the MES.$2H_2O$ to the lower temperature side and the appearance of a new peak at 53°C. At 19.5% moisture content a peak that apparently corresponds to the phase transition of the MES.$5H_2O$ crystal was observed. At 19.5 to 32.6% moisture content a shift of the endothermic peak corresponding to the phase transition of MES.$5H_2O$ crystals to the lower temperature side occurred, accompanied by a peak at 42°C, which was assigned to the phase transition of MES.$10H_2O$ crystals. The transition temperature change in this moisture content range in all cases was attributed to the mixture of MES.$5H_2O$ and MES.$10H_2O$ crystals.

Based on these findings, an overall phase diagram of the MES/water system is proposed in Figure 6.8, in which four different hydrated crystals (anhydrate, 2 mole hydrate, 5 mole hydrate and 10 mole hydrate), two liquid crystals (lamellar and hexagonal), a micelle solution and a monomer solution are presented.

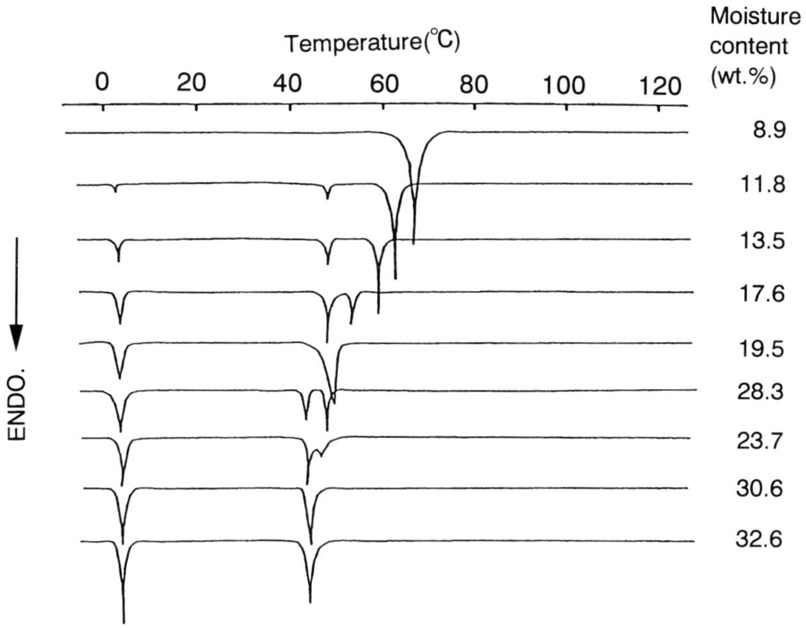

Figure 6.7 DSC curves for MES solids with moisture contents from 8.9 to 32.6 wt%.

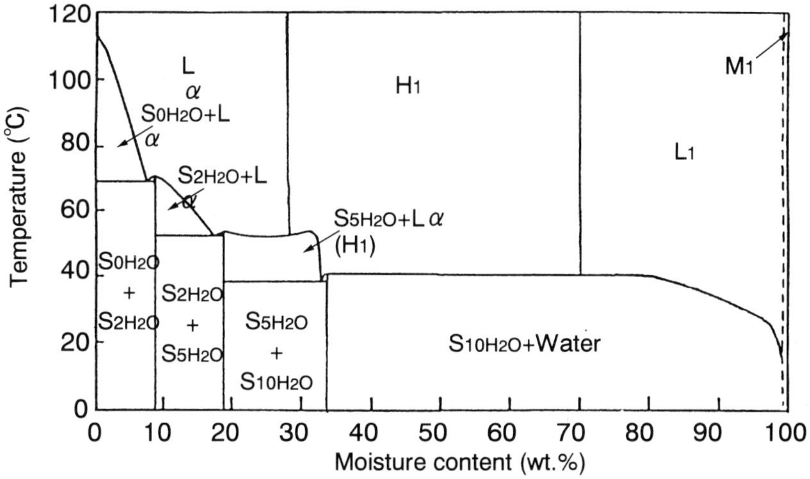

Figure 6.8 Phase diagram of the MES–water system. S_0H_2O = anhydrous MES; S_2H_2O = MES.2H_2O; S_5H_2O = MES.5H_2O; $S_{10}H_2O$ = MES.10H_2O; Lα = lamellar liquid crystalline phase; H1 = hexagonal liquid crystalline phase; L1 = micellar solution phase; M1 = monomer solution phase.

MES crystals have several different hydrations and this can be studied in relation to the atmosphere. The moisture content and X-ray diffraction spectra of MES solid samples with 40% initial moisture content were determined at a constant relative humidity of 10% and ambient temperature (Figure 6.9). The moisture content initially decreased rapidly for 5 h and then remained constant at 32 wt% up to 20 h. There was then a gradual decrease from 20 to 50 h, reaching 9 wt%, corresponding to the phase transition of MES.2H_2O. The moisture content then fell to 7 wt% and remained constant. The X-ray diffraction patterns indicated essentially the same results at 40 to 10 wt% moisture content, where two peaks of 4.1 and 4.2 Å appeared. Finally, at 7 wt% moisture content a single peak of 4.25 Å was observed.

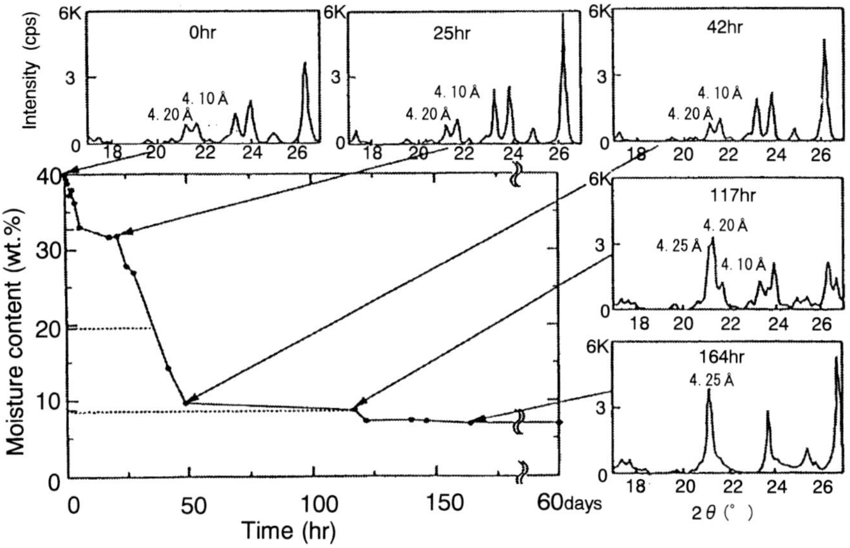

Figure 6.9 Time dependence of moisture content for the hydrated MES solid. Conditions: initial moisture content 40 wt%, 10% relative humidity, room temperature.

MES crystals in the dry state were also studied in a humid atmosphere. MES solids with initial 0.2 wt% moisture content were maintained under 80% relative humidity at 50°C. Phase changes are shown in Figure 6.10.

The moisture content gradually increased up to 20 h and then leveled off near 25 h, where it became 8.6 wt%, corresponding to MES.2H_2O crystals. X-ray diffraction showed two peaks of 4.1 and 4.2 Å at 0.2 to 4 wt% moisture content, and then a single peak of 4.25 Å at a moisture content less than 6 wt%.

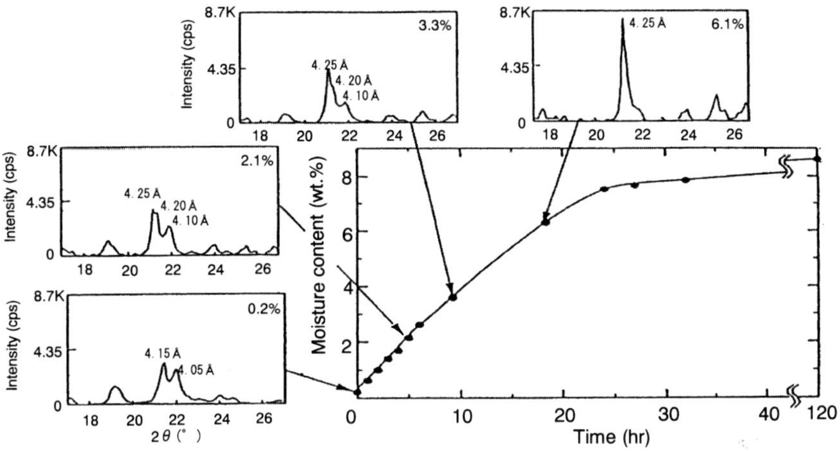

Figure 6.10 Time dependence of moisture content for the dry MES solid. Conditions: initial moisture content 0.2 wt%, 80% relative humidity, 50°C.

It follows from the above that MES crystals may easily take on the MES.2H_2O state at ambient temperature and they are most stable in this state. The phase transition temperatures and enthalpy of MES hydrated crystals in Table 6.3 indicate that the 2 mole hydrate is possibly the most stable form. The enthalpy of MES.2H_2O is the largest of all the hydrated MES crystals, suggesting that dense molecular packing to gives rise to maximum stability.

Table 6.3 Phase transition temperatures and enthalpies of anhydrous and hydrated MES crystals

MES crystal	T_c (°C)	ΔH_2(kJ/mol)
Anhydrous MES	112	22.1
MES.2H_2O	70	67.5
MES.5H_2O	53	39.4
MES.10H_2O	42	46.1

6.5 Application in laundry detergents

The choice of surfactants for laundry detergents is based on factors such as performance, manufacturing technology, social considerations and demands for feedstock. In Japan, in response to environmental concern in the 1970s, zeolite replaced polyphosphate as a detergent builder and new surfactants from natural fats and oils attracted much attention. MESs, which appeared in 1991, were considered suitable for a compact detergent in Japan.

The practical detergency and the surface activity of MESs are discussed in the following sections, along with formulations and the manufacturing process.

6.5.1 Detergency

The detergency of MESs for different alkyl chain lengths has been evaluated for comparison with LASs and ASs (C_{12}) at low temperatures and water hardness [24]. Detergency was studied using the Terg-O-Tometer (US Testing) and artificial soiled swatches as fabrics at 120 rpm, 25°C and a 10 min wash plus two rinses of 3 min each. Sodium carbonate (135 ppm) and sodium silicate (35 ppm) were used in each washing test. Water hardness was controlled to 3°DH. Detergency was determined based on a swatch reflectance before and after washing and the Kubelka–Munk equation [25]. The swatches were prepared by soaking unsoiled swatches in an aqueous dispersion containing oil, protein and mineral components, followed by drying in air [25].

Detergency test results on three different surfactants as a function of concentration are shown in Figure 6.11. C_{16}MES and C_{18}MES showed better detergency than LASs or, in particular, ASs and the best results were obtained at low surfactant concentrations. C_{16}MES gave the best detergency, followed by C_{18}MES and then C_{14}MES. Detergency followed the order C_{16}MES ≥ C_{18}MES > LAS > AS, C_{14}MES.

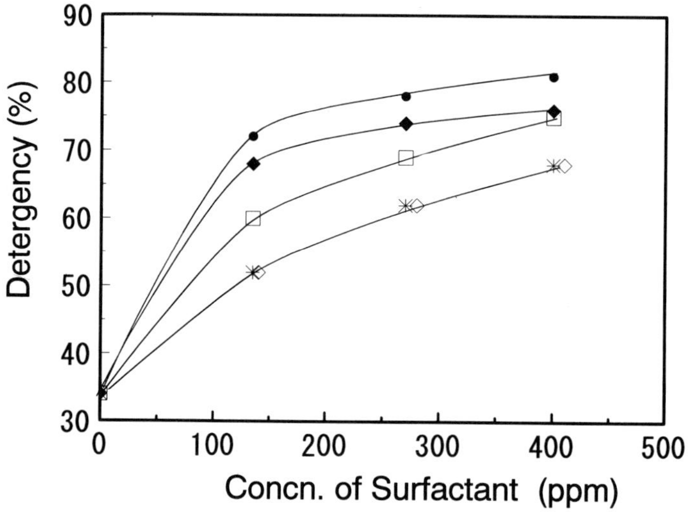

Figure 6.11 Relationship between detergency and concentration of surfactant. Conditions: Terg-O-Tometer, artificial soil (cotton), 25°C, water hardness 54 ppm ($CaCO_3$). Key: C_{18}MES ◆, C_{16}MES ●, C_{14}MES *, LAS □, AS ◊.

The effects of temperature (Figure 6.12) and liquor ratio (Figure 6.13) on detergency have been evaluated [24]. Detergency decreases with reduction in temperature and results for MESs were best at low temperatures and low liquor ratio.

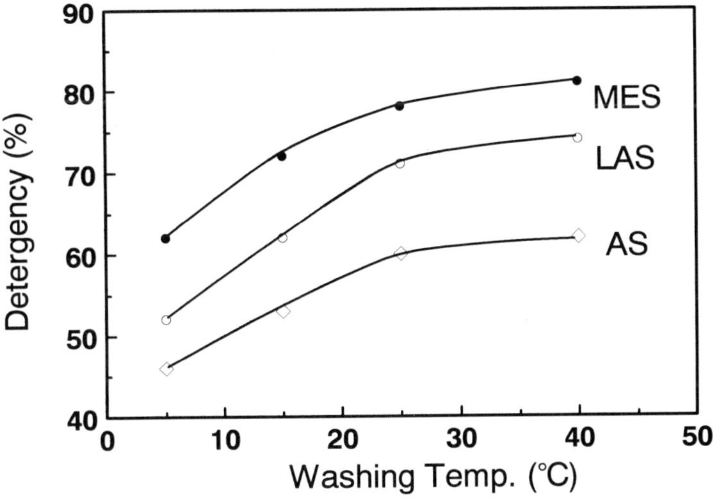

Figure 6.12 Effect of washing temperature on detergency. Conditions: Terg-O-Tometer, artificial soil, liquor ratio 30, water hardness 54 ppm ($CaCO_3$), surfactant 270 ppm, sodium carbonate 135 ppm, sodium silicate 135 ppm.

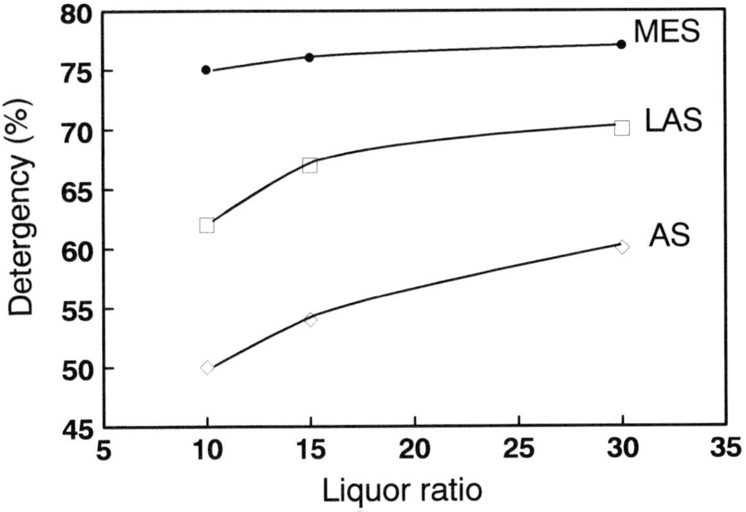

Figure 6.13 Effect of liquor ratio on detergency. Conditions: Terg-O-Tometer, artificial soil, 25°C, water hardness 54 ppm ($CaCO_3$), surfactant 270 ppm, sodium carbonate 135 ppm, sodium silicate 135 ppm.

The surface-active properties of the three surfactants were examined to determine differences in performance. Adsorption onto particle soil, dispersion of clay, emulsification of oily soil and the zeta potentials of oil droplets in the emulsion system were measured for C_{14-16}MES, LAS and AS solutions as a function of surfactant concentration. The MES surfactants gave the best results in all cases, as shown in Figures 6.14 (adsorption) and 6.15 (emulsification) [24].

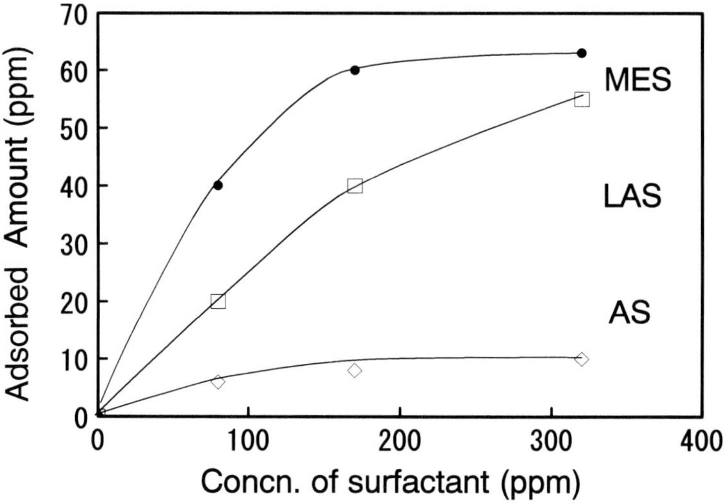

Figure 6.14 Adsorption of surfactants onto particle soil. Conditions: clay 0.4%, 25°C, sodium carbonate 270 ppm.

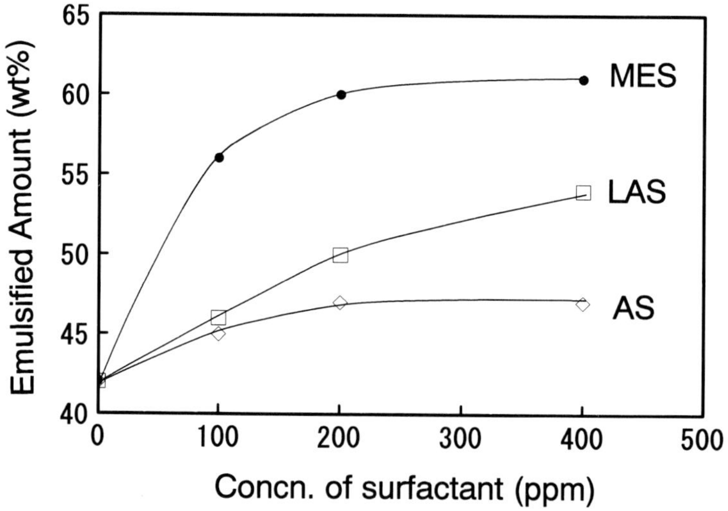

Figure 6.15 Emulsification of oily soil by surfactants. Conditions: artificially oily soil 0.2%, 25°C, sodium sulfate 270 ppm.

The adsorption of each surfactant on clay particles was determined for an aqueous dispersion prepared by agitating red yellow diluvium for 3 h in the presence of each surfactant. In the emulsifying tests emulsified oil was proportional to surfactant concentration.

6.5.2 Solubilization

MESs have superior solubilization capacity, which is essential for characterizing surfactants. MESs were found to be highly capable of solubilizing oleic acid, which is usually present abundantly in natural soil. The solubilization capacity for LASs, ASs and C_{14-16}MES was evaluated using oleic acid as a polar oil solubilizate and a non-polar *n*-octane [24]. For this purpose a definite weight of oleic acid was introduced to the surfactant solution, followed by shaking for 24 h at 30°C, and standing for 12 h for visual assessment of the transparency. As can be seen from Figure 6.16, C_{14-16}MES showed the best solubilization for oleic acid. For the non-polar oil, the MES and LAS samples were essentially the same. Light scattering indicated the MESs micelle diameter to be twice that of LASs, as indicated in Figure 6.17.

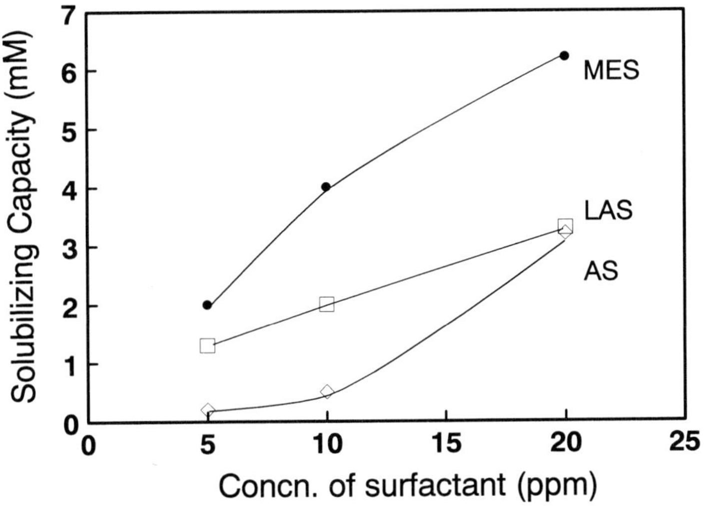

Figure 6.16 Solubilizing capacity of surfactants for oleic acid.

Fujiwara [26] investigated the solubilization of MES and obtained the following results:

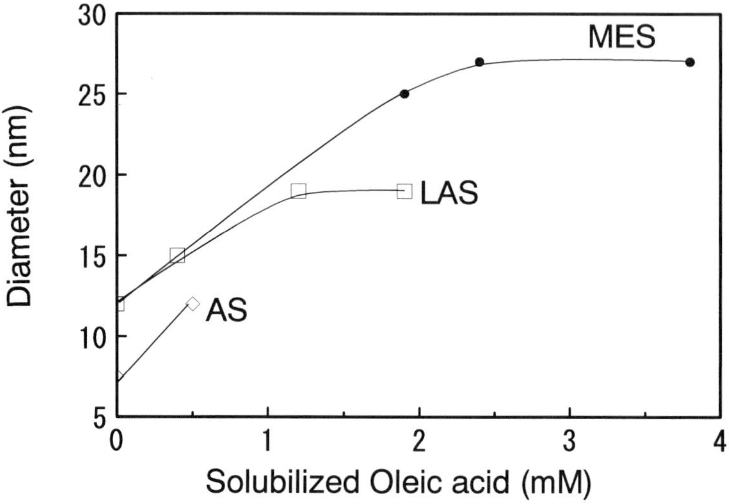

Figure 6.17 Relationship between solubilized micelle diameter and concentration of oleic acid.

- MES has a larger solubilization capacity than LAS
- capacity depends on the number of micelles and degree of solubilization per micelle
- miceller weight initially decreases for a small amount of solubilizate and then increases with increasing solubilizate amount
- the solubilization process is accompanied by the reconstitution of micelles.

6.5.3 Enzyme stability

Enzymes are essential components of detergents. Surfactants have significant influence on enzyme activity in detergents. A stable enzyme and surfactant system is desirable for optimal enzyme functioning in washing. Non-ionic surfactants have little effect on enzyme activity, whereas cationic surfactants have considerable effect. Anionic surfactants exert an intermediate effect, the extent depending on the surfactant. The effects of three anionic surfactants on enzyme activity were examined using protease as the detergent enzyme [27]. Activity stability in a washing liquor was assessed at low temperature, low water hardness and weak alkaline pH. The results are shown in Figure 6.18. Enzyme activity decreased with time for all the surfactants. In MES solution only a slight decrease in activity was noted, whereas that for LAS and AS solutions was significant; residual activity for either solution was below 50% after standing for 2 h at 25°C. The inhibition of enzyme activity due to surfactants is thought to be caused by adsorption of the surfactant onto the enzyme, leading to denaturation of the enzyme protein.

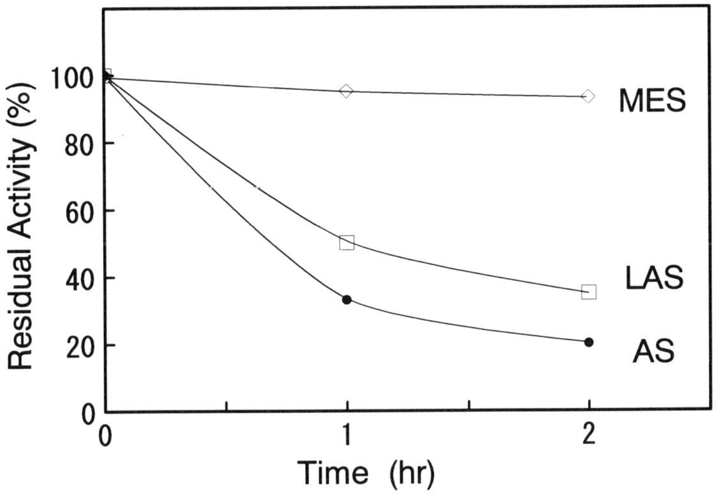

Figure 6.18 Effect of surfactants on protease activity. Conditions: surfactant 300 ppm, protease 0.008 AU/l, pH 10.5, 40°C.

6.5.4 Formulations

The best feature of MESs is that its surfactant content can be reduced without compromising performance. Formulations containing MESs are listed in Table 6.4 [28]. Compared with the popular formulation of LAS/AS or LAS/AOS, those of MESs allow for sufficient detergency with low surfactant content, with no need for additional sequestering or alkaline builders.

Table 6.4 Formulations of MES-based detergents

Component	LAS-based	MES-based	MES/LAS-based
Surfactant	30–40 (LAS/AOS, LAS/AS)	30–40	30–40
Builder	15–25	15–25	15–25
Alkali	15–25	15–25	15–25
Enzyme	+	+	+
FWA	+	+	+

FWA = fluorescent whitening agent.

Detergent powders containing MESs as the main component have various technical problems, such as a low dissolving rate in low temperature water, possibly due to the high Krafft point and high crystallinity of MESs. Ester bonds in the MES molecule may easily undergo hydrolysis with consequent formation of the disodium salt, which has low surface activity. The hydrolysis of MESs in detergent powder may occur on contact with alkaline components.

A new surfactant system, which uses LAS with MES, has been established and good results obtained from it, as shown in Table 6.4 [28]. This system not only eliminates the disadvantages of MESs but enhances washing detergency at low temperatures.

The surfactant properties of the MES/LAS blend were investigated physicochemically by Krafft points determined from the cross-temperature point of the solubility line and the CMC line against temperature (Figure 6.19). The value of the Krafft point decreased with increasing LAS. The blend was also studied by DSC using a surfactant powder with 5% moisture content.

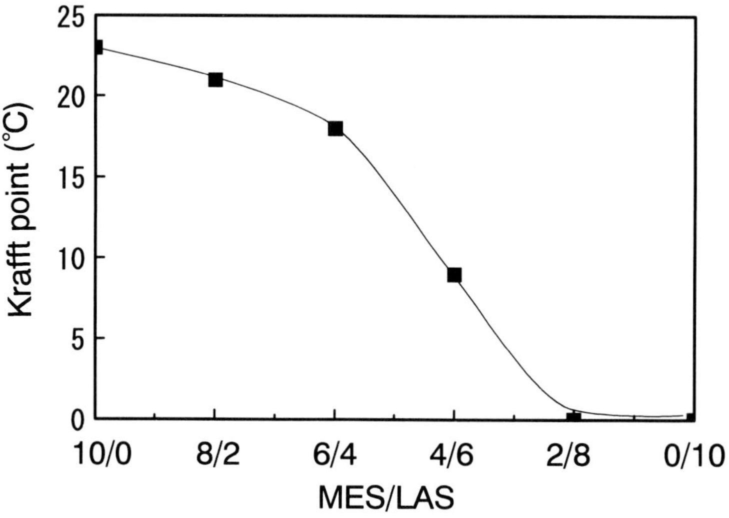

Figure 6.19 Krafft point of MES/LAS mixture.

MES powder is a 2 mole hydrate, judging from moisture content. Figure 6.20 shows the DSC pattern for C_{14-16}MES. An endothermic peak can be seen at 39°C, lower than that of C_{16}MES (see Figure 6.5). For LAS powder no endothermic peak was observed. MES/LAS in a one-to-one ratio clearly showed a smaller DSC peak, indicating that the MES powder no longer had high crystallinity. DSC and Krafft point data suggest that MESs should be mixed with LASs at a molecular level to reduce the DSC endothermic peak and fall-off of the Krafft points. This mixing would also enhance detergency, which was assessed under the usual conditions using artificial swatches (Figure 6.21). Detergency was found to increase with the ratio of MES to LAS at 25°C. At 15°C, the detergency was a maximum at a one-to-one blend ratio.

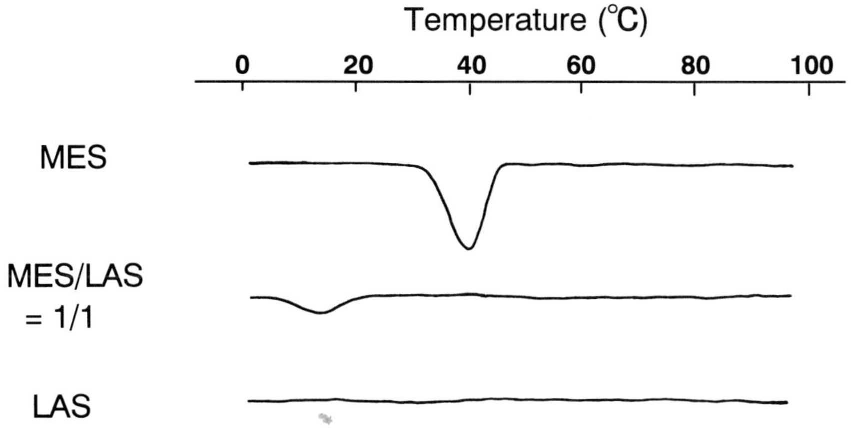

Figure 6.20 DSC profile of MES/LAS mixture. Sample: surfactant powder, water content 5%, DSC condition 1°C/min.

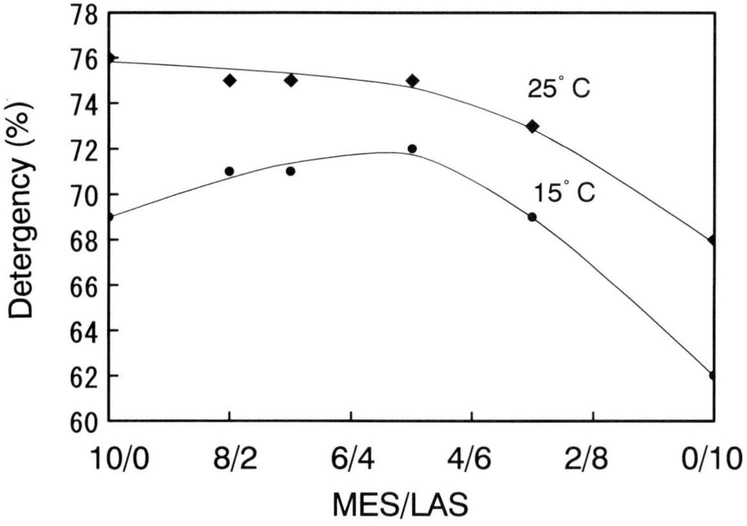

Figure 6.21 Detergency of MES/LAS mixture. Conditions: Terg-O-Tometer, artificial soil, (cotton), water hardness 54 ppm ($CaCO_3$), surfactant (MES + LAS) 200 ppm, sodium carbonate 200 ppm, zeolite 150 ppm.

6.5.5 Manufacturing process [28]

Powder detergents are generally produced by drying an aqueous slurry of raw material in a spray tower. The most serious problem with MES detergents is hydrolysis during the drying process, the high temperature of which promotes the hydrolysis of MES ester bonds to readily produce the disodium salt. High pH and high temperature should therefore be avoided in MES detergent production (Figure 6.22).

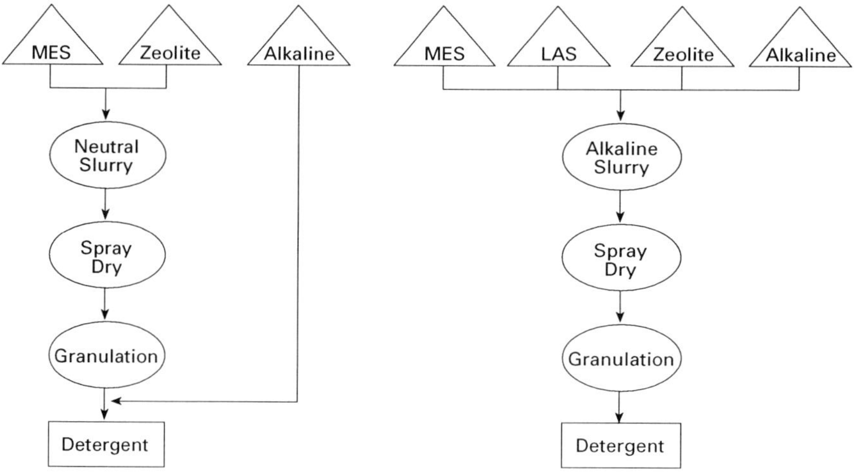

Figure 6.22 Process flow of MES/LAS detergents.

In the MES/LAS blend, the spraying temperature may be maintained at much the same as that of the LAS detergent because of its good heat stability. In addition, moisture is rapidly discharged during drying, with loss of high MES crystallinity. Thus, even in the presence of alkaline builder hydrolysis is kept low in the MES/LAS blend. MESs were formerly considered to be complicated surfactants because of the difficulty in manufacturing them, but now they are considered to be about the same as conventional LAS surfactants because the two may be used together as a blend and because of technical advances.

6.6 Biodegradability

The biodegradability of MESs has been reported previously [29]. Masuda *et al.* reported studies which used a different examination method and proposed a pathway of MES biodegradation by microorganisms [30, 31]. These studies involved a shaking culture method, river die-away test and biochemical oxygen demand measurement (Japanese MITI test). Biodegradation was monitored by methylene blue active substances (MBAS), dissolved organic carbon (DOC) and biochemical oxygen demand (BOD). Results from the shaking culture method are given in Figure 6.23. It can be seen that MESs, as well as AOSs, lose more than 90% of methylene blue (MB) activity in a day and 100% within two days. More than 80% of the organic carbon in the surfactants disappears in two days. LASs lost MB activity in five days and more than 40% of the dissolved organic carbon still remained after 15 days. MESs are thus shown to be readily biodegradable substances from the viewpoint of both primary and ultimate degradation under shaking culture conditions.

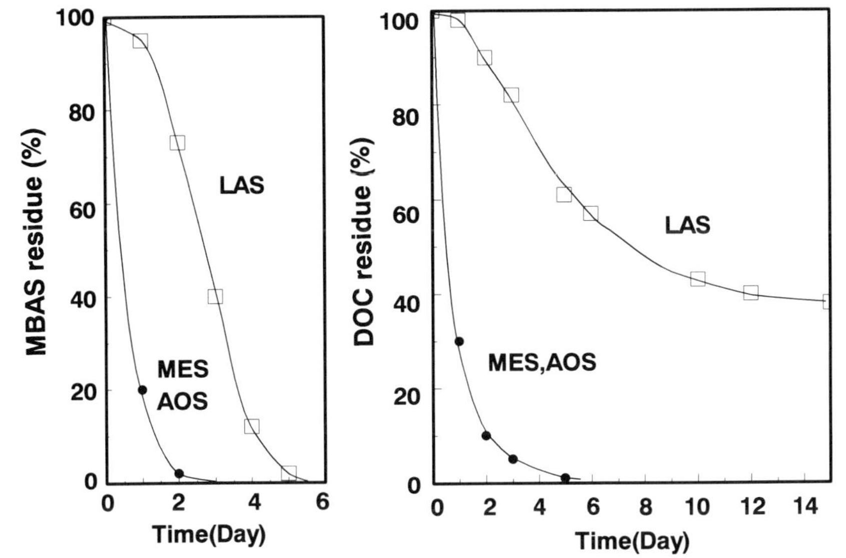

Figure 6.23 Biodegradation of surfactants by shake culture method.

The river die-away test, a biodegradation test using actual river water, was conducted on four different surfactants. Figure 6.24 shows that MES and AOS surfactants quickly and easily undergo biodegradation, with total elimination of

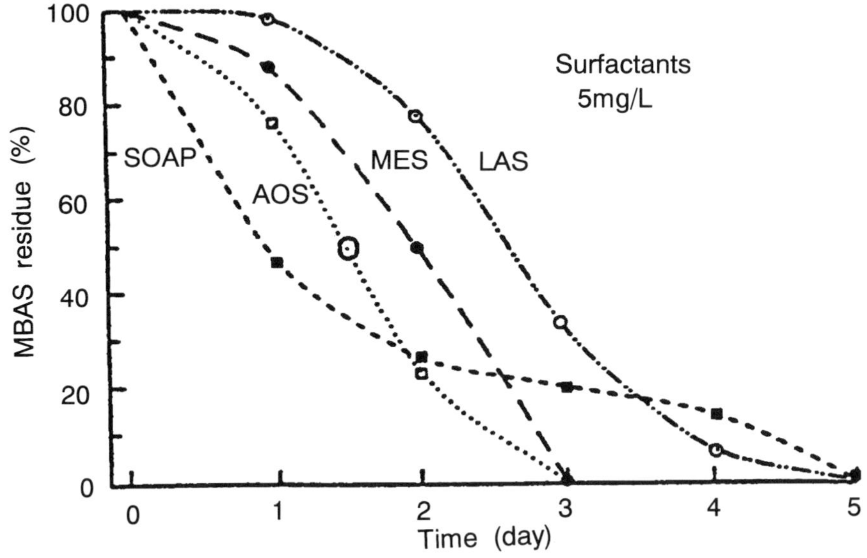

Figure 6.24 Biodegradation of surfactants in river water (river die-away test) detected by MBAS. Surfactants 5 mg/l.

MB activity, within three days. For soap, five days were required for loss of MB activity, suggesting that soap causes some kind of precipitation in river water whereby degradation is delayed.

The results of the BOD/TOD (total oxygen demand) measurements in the MITI test are shown in Figure 6.25. These results indicate that the biodegradation of MESs starts quickly and proceeds rapidly in the early stage at each surfactant concentration. For LASs degradation starts later and is in proportion to the increase in surfactant concentration. Results also indicate that both MES and LAS surfactants undergo ultimate biodegradation.

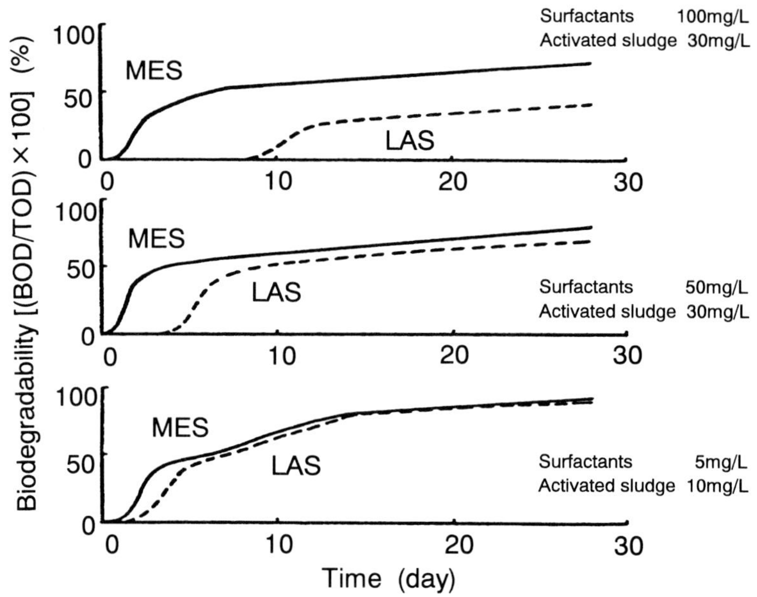

Figure 6.25 Biodegradation of MESs and LASs detected by BOD.

Masuda [29] determined the pathway of MES biodegradation based on change in the chemical structure of MESs in the MITI test. The biodegradation test was carried out for C_{14}MES using 50 mg/l surfactant and 30 mg/l activated sludge, and the BOD was measured. Solutions at each step of biodegradation were sampled, extracted and then analyzed by NMR, IR and HPLC. The results of HPLC analysis to detect the inorganic sulfate are shown in Figure 6.26. In the early stages of degradation no inorganic sulfate was detected. Inorganic sulfate was found at the halfway stage and it subsequently reached 71% of the total.

Analysis of degradation intermediates indicated the biodegradation pathway shown in Figure 6.27, in which there is a microbial attack on the surfactant structure; ω-oxidation produces a carboxyl group and with β-oxidation two

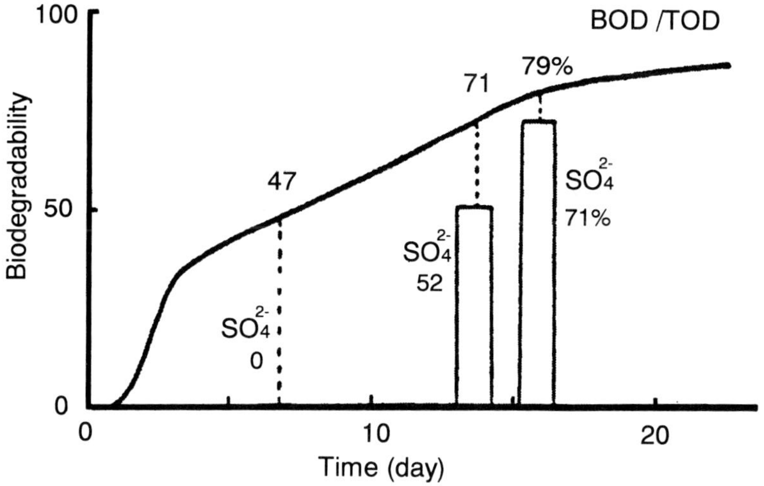

Figure 6.26 SO_4^{2-} ion release during biodegradation of MESs.

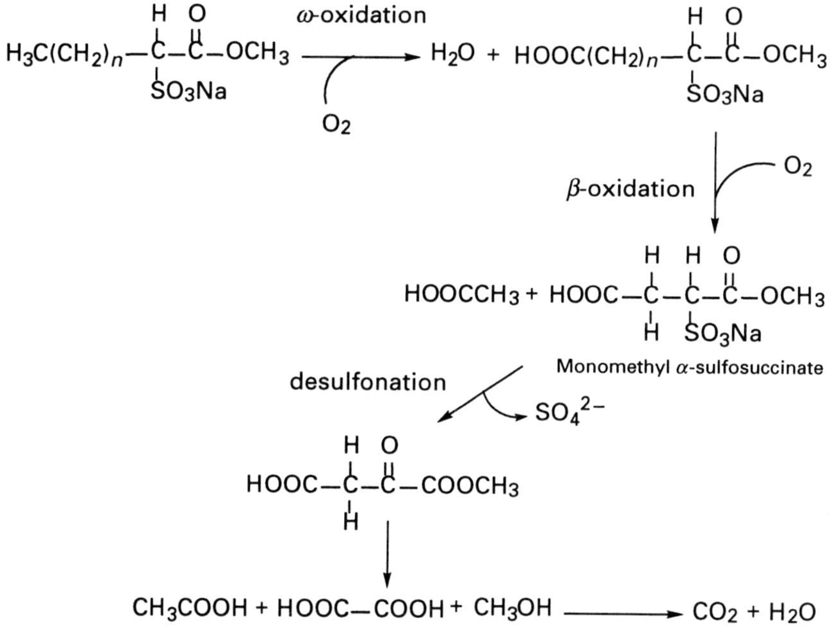

Figure 6.27 Biodegradation pathway of MES.

carbons are simultaneously removed. The desulfonation of the temporary intermediate monomethyl alpha-sulfosuccinate then occurs following scission of the methyl ester bond. The monomethyl alpha-sulfosuccinate eventually undergoes degradation to produce carbon dioxide.

Acknowledgements

The author expresses sincere appreciation to Dr Yuji Yoneyama for helping with this work and to Dr Haruo Yoshimura and Mr Toshiyuki Watanabe for their critical review of the manuscript.

References

[1] Stirton, A.J., Bistline, Jr, R.G., Weil, J.K. and Maurer,J. (1962) Sodium salts of alkyl esters of alpha-sulfo fatty acids: wetting, lime soap dispersion and related properties. *Journal of the American Oil Chemists' Society*, **39** 128-31.

[2] Boucher, E.A., Grinchuk, T.M. and Zettlemoyer, A.C. (1968) Surface activity of sodium salts of alpha-sulfo fatty esters: the air–water interface. *Journal of the American Oil Chemists' Society*, **45** 49-52.

[3] Okumura, O., Sakatani, T. and Yamane, I. (1976) Mechanism of sulfonation of fatty acid esters with sulfur trioxide and properties of alpha-sulfo fatty acid esters, in *Proceedings of the 7th Comité International des Derives Tensio-Actis*, vol. 1, Moscow, pp. 225-34.

[4] Stirton, A.J.,Weil, J.K. and R.G. Bistline, Jr, R.G. (1954) Surface-active properties of salts of alpha-sulfonated acids and esters. *Journal of the American Oil Chemists' Society*, **31** 13-16.

[5] Schambil, F. and Schwuger, M.J. (1990) Physico-chemical properties of alpha-sulpho fatty acid methyl esters and alpha-sulpho fatty acid di-salts. *Tenside Surfactants Detergents*, **27** (6) 380-29.

[6] Maurer, E.W., Weil, J.K. and Linfield, W.M. (1977) The biodegradation of esters of alpha-sulfo fatty acids. *Journal of the American Oil Chemists' Society*, **54** 582-84.

[7] Steber, J. and Wierich, D. (1989) The environmental fate of fatty acid alpha-sulfomethyl esters. *Tenside Surfactants Detergents*, **26** 406-11.

[8] Stirton, A.J. (1968) Alpha-sulfo fatty acids and derivatives: synthesis, properties and use. *Journal of the American Oil Chemists' Society*, **39** 490-96.

[9] Weil, J.K., Bistline, Jr., R.G. and Stirton, A.J. (1953) Sodium salts of alkyl alpha-sulfo-palmitate and stearates. *Journal of the American Oil Chemists' Society*, **75** 4859-60.

[10] Stein, W. and Baumann, H. (1975) Alpha-sulfonated fatty acids and esters: manufacturing process, properties and applications. *Journal of the American Oil Chemists' Society*, **52** 323-29.

[11] Kapur, B.L., Solomon, J.M. and Bluestein, B.R. (1978) Summary of the technology for the manufacture of higher alpha-sulfo fatty acid esters. *Journal of the American Oil Chemists' Society*, **55** 549-57.

[12] Yamane, I. and Miyawaki, Y. (1989) Manufacturing process of alpha-sulphomethyl esters and their application to detergents, in *Proceedings of 1989 International Palm Oil Development Conference: Chemistry, Technology, and Marketing*, Malaysia, pp. 132-41.

[13] Drozd, J.C. and Desai, D.D. (1991) Liquid laundry detergents based on soap and alpha-sulfo methyl esters. *Journal of the American Oil Chemists' Society*, **68** 59-62.

[14] Smith, N.R. (1989) Alpha-sulfo methyl esters: a new alternative. *Soap Cosmetics Chemical Specialties*, April, 48-57, 130, 134, 136.

[15] Fujiwara, M., Miyake, M. and Abe, Y. (1993) Colloidal properties of alpha-sulfonated fatty acid methyl esters and their applicability in hard water. *Colloid Polymer Science*, **271** 780-85.
[16] Hato, M. and Shinoda, K. (1973) The solubilities, critical micelle concentrations, and Krafft points of bivalent metal alkyl sulfates. *Bulletin of the Chemical Society of Japan*, **46** 3889-90.
[17] Shinoda, K., Maekawa, M. and Shibata, M. (1986) Ionic surfactants soluble in hard water and hydrocarbons: behavior of organized surfactant solutions as a function of the hydrophilic–lipophilic balance. *Journal of Physical Chemistry*, **90** 1228-30.
[18] Shinoda, K. (1963) The formation of micelles, in *Colloidal Surfactants*, Academic Press, New York, pp. 55-57.
[19] Hayashi, S. and Ikeda, S. (1980) Micelle size and shape of sodium dodecyl sulfate in concentrated NaCl solution. *Journal of Physical Chemistry*, **84** 744.
[20] Stellner, K.L. and Scamehorn, J.F. (1989) Hardness tolerance of anionic surfactant solutions 1. Anionic surfactant with added monovalent electrolyte. *Langmuir*, **5** 70-77.
[21] Kekicheff, P., Grabielle-Madelmont, C. and Ollivon, M. (1989) Phase diagram of sodium dodecyl sulfate–water system: a calorimetric study. *Journal of Colloid Interface Science*, **131** (1) 112-32.
[22] Fujiwara, M., Okano, T., Amano, H., Asano, H. and Ohbu, K. (1997) Phase diagram of alpha-sulfonated palmitic acid methyl ester sodium salt/water system, in *Proceedings of the 4th CESIO International Surfactant Congress*, Section D, Barcelona, pp. 201-10.
[23] Wittman, J.C., and Manley, R. St. J. (1977) Polymer–monomer binary mixture. I. Eutetic and epitaxial crystallization in poly(e-caprolactone)-trioxane mixtures. *Journal of Polymer Science, Part B Polymer Physics*, **15** 1089-1100.
[24] Satsuki, T., Umehara, K. and Yoneyama, Y. (1992) Performance and physicochemical properties of alpha-sulfo fatty acid methyl esters. *Journal of the American Oil Chemists' Society*, **69** 672-77.
[25] Okumura, O., Tokuyama, K., Sakatani, K. and Tsuruta, T. (1980) *Journal of the Japanese Oil Chemists' Society* (*Yukagaku*), **30** 432-41.
[26] Fujiwara, M., Kaneko, Y. and Ohbu, K. (1995) Light scattering study on the micellar systems solubilizing a fatty acid. *Colloid Polymer Science*, **273** 1055-59.
[27] Satsuki, T. (1992) Application of MES in detergents. *Inform*, **3** 1099-1108.
[28] Satsuki, T. (1986) Methyl ester sulfonates: a surfactant based on natural fats, in *Proceeding of the 3rd World Conference on Detergents*, Montreux, pp. 135-40
[29] Miura, K. (1991) Performances of fatty acid alpha-sulfomethyl esters (3). Biodegradation, in *3rd Annual Meeting of Home Economics*, Tokyo.
[30] Masuda, M., Odake, H., Miura, K. and Oba, K. (1993) Biodegradation of 2-sulfonatofatty acid methyl ester I. *Journal of the Japanese Oil Chemists' Society* (*Yukagaku*), **42** 643-47.
[31] Masuda, M., Odake, H., Miura, K., Ito, K., Yamada, K., and Oba, K. (1993) Biodegradation of 2-sulfonatofatty acid methyl ester II. *Journal of the Japanese Oil Chemists' Society* (*Yukagaku*), **42** 905-909.

7 Fluorocarbon–hydrocarbon hybrid surfactants

Masahiko Abe and Norio Yoshino

7.1 Introduction

Fluorocarbon compounds are well known to exhibit water and oil repellency, high lubricity, and thermal and chemical stability. Fluorocarbon surfactants containing a fluorocarbon chain as the hydrophobic group display peculiar abilities, such as thermal and chemical stability, high surface activity, high surface modifiability and strong micelle forming ability [1, 2]. Extremely heterogeneous micelles are reported to form in mixed solutions of fluorocarbon surfactant and hydrocarbon surfactant [3–5]. Hybrid surfactants are composed of a hydrocarbon chain and a fluorocarbon chain attached to the same hydrophilic head group. In 1992, a new series of hybrid anionic surfactants, for example $C_7F_{15}CH(OSO_3Na)C_7H_{15}$, was first synthesized by Wen Guo *et al.* [6, 7] However, the surfactants were found to slowly hydrolyze in air through the adsorption of moisture. They therefore had to be stored in a desiccator at –25°C. They also hydrolyzed slowly in aqueous solution at room temperature and hence all physical measurements had to be made within 20 h after sample preparation. These results, however, provide good indicators for the preparation of hydrolysis-resistant hybrid surfactants. Thus, six novel hybrid surfactants having both a fluorocarbon chain and a hydrocarbon chain with an aromatic ring have been synthesized (Figure 7.1) [8]. The general formula for these compounds is FC*m*–HC*n*. These surfactants are stable in the presence of moisture owing to the resonance between the benzene ring and the carbonyl group and to the bonding structure of CSO_3Na at the hydrophilic segment, which is resistant to hydrolysis.

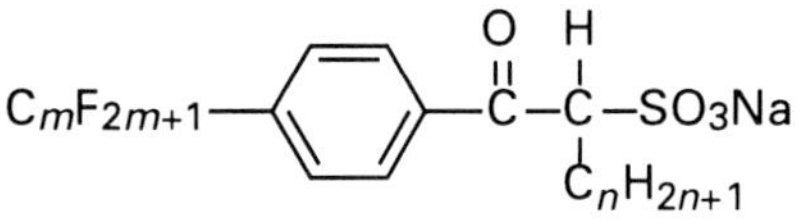

$n = 2, 4, 6 \quad m = 4, 6$

Figure 7.1 Chemical structure of fluorocarbon–hydrocarbon hybrid surfactants.

7.2 Surface chemical properties

The syntheses and purifications of these new hybrid surfactants are described in reference [8].

Table 7.1 shows the surface chemical properties of aqueous FC*m*–HC*n* solutions. The Krafft points (K_p) of FC4–HC2, FC4–HC4 and FC6–HC4 are below 0°C, which indicates their superior solubility in water. K_p tends to increase with an increase in fluorocarbon and/or hydrocarbon chain length.

Table 7.1 also shows the critical micelle concentrations (CMC) and surface tension values at the critical micelle concentration (γ_{CMC}) of aqueous FC*m*–HC*n* solutions. The γ_{CMC} values are 24.0 mN/m (FC4–HC2), 22.5 mN/m (FC4–HC4), 18.9 mN/m (FC4–HC6), 21.5 mN/m (FC6–HC2), 20.4 mN/m (FC6–HC4), and 16.2 mN/m (FC6–HC6, 50°C). These values indicate that the FC*m*–HC*n*s are superior in surface activity to other surfactants, and FC4–HC6 is the most effective in lowering surface tension at 25°C. Moreover, γ_{CMC} decreases with an increase in fluorocarbon and/or hydrocarbon chain length.

The interfacial tensions between *n*-octane and the surfactant solutions at CMCs are 9.6 mN/m (FC4–HC2), 6.2 mN/m (FC4–HC4), 2.7 mN/m (FC4–HC6), 7.7 mN/m (FC6–HC2), 6.4 mN/m (FC6–HC4) and 3.4 mN/m (FC6–HC6, 50°C), and decrease with an increase in fluorocarbon and/or hydrocarbon chain length. FC*m*–HC*n*s also show lower interfacial tension between water and liquid fluorocarbon (not shown in Table 7.1).

It should be noted that aqueous FC*m*–HC*n* solutions can float on liquid hydrocarbons such as benzene, toluene, xylene, cyclohexane, dodecane and decane, as shown in Figure 7.2 (for CF4–HC6) [9].

The following equation must hold in order for an aqueous surfactant solution to spread on the oil surface [10].

$$\gamma_{a/o} > \gamma_{a/w} + \gamma_{o/w} \tag{7.1}$$

where $\gamma_{a/o}$ and $\gamma_{a/w}$ are the surface tensions of the oil and the aqueous surfactant solution, respectively, and $\gamma_{o/w}$ is the oil/solution interfacial tension. The values of $\gamma_{a/o}$ for FC*m*–HC*n*s are larger than the sum of $\gamma_{a/w}$ and $\gamma_{o/w}$ (Table 7.1), and this makes it possible for the surfactant solutions to spread on the oil.

Table 7.1 Surface chemical properties of FC*m*–HC*n*

	m=4			*m*=6		
	n=2	*n*=4	*n*=6	*n*=2	*n*=4	*n*=6
Molecular weight	468.3	496.3	524.4	568.3	596.3	624.4
Krafft point (°C)	<0	<0	20	<0	15	48
CMC (mol/l)	8.2×10^{-3}	3.5×10^{-3}	1.2×10^{-3}	8.3×10^{-4}	2.3×10^{-4}	5.5×10^{-5}
γ_{CMC} (mN/m)	24.0	22.5	18.9	21.5	20.4	16.2
Air/water molecular occupied area (Å^2/molec.)	81.3	89.0	106.0	89.4	104.7	98.3
Octane/water interfacial tension (mN/m)	9.6	6.2	2.7	7.7	6.4	3.4
Octane/water molecular occupied area (Å^2/molec.)	82.2	95.5	122.7	116.5	129.7	107.3
Aggregation number of micelle	14	18	25	17	22	21
Air/water Szyszkowski constant (mN/m)	23.3	21.3	17.9	21.2	18.1	20.9
Air/water Langmuir constant (l/mol)	1.4×10^{4}	5.2×10^{4}	7.0×10^{5}	2.9×10^{5}	3.2×10^{6}	5.5×10^{6}
Air/water ΔG_{ad} (kJ/mol)	−29.4	−32.4	−38.5	−36.7	−42.2	−47.6
Octane/water Szyszkowski constant (mN/m)	23.1	19.9	15.5	16.3	14.6	19.2
Octane/water Langmuir constant (l/mol)	7.3×10^{3}	4.3×10^{4}	8.9×10^{5}	2.7×10^{5}	3.2×10^{6}	4.6×10^{6}
Octane/water ΔG_{ad} (kJ/mol)	−27.8	−31.8	−38.7	−35.8	−41.7	−46.9

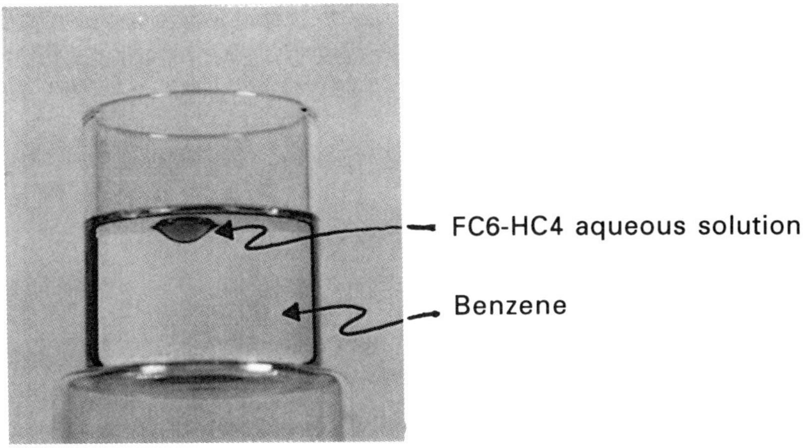

Figure 7.2 Photograph of an aqueous surfactant solution drop on benzene. Drop: aqueous FC4–HC6 solution (5 mM). Solvent: benzene.

The aggregation numbers, *N*, of FC*m*–HC*n* micelles at their CMCs are smaller than those of typical ionic hydrocarbon surfactants (about 50–70), presumably because of the shorter hydrophobic chain lengths of FC*m*–HC*n*s. *N* increases with an increase in fluorocarbon or hydrocarbon chain length, except for FC6–HC6, which is estimated at a higher temperature. Moreover, FC*m*–HC*n*s are found to have a large area occupied by a molecule, probably because of the existence of a phenylene group and the repulsion between fluorocarbon chain and hydrocarbon chain in each molecule, as shown in Table 7.1. Table 7.1 also gives the values of the Szyszkowski constant, Langmuir constant and standard free energy for the adsorption of these surfactants.

Perfluoropolyethers are well known to be immiscible with organic solvents except fluorine-containing solvents, such as chlorofluorocarbons (CFCs). However, a ternary system, octane/water/perfluoropolyether oil, each component of which is immiscible with the others, can be emulsified under ultrasonic irradiation using a small amount of the hybrid surfactant, as shown in Figure 7.3 (for CF6–CH6). The hybrid surfactants can easily emulsify not only two-component systems such as perfluoropolyether oil/water, but also three-component systems such as hydrocarbon/water/perfluoropolyether oil. Utilization of the hybrid surfactant should enable the development of fluoro-type paints using water as the dilution solvent.

Figure 7.3 Ternary system, octane/water/perfluoropolyether oil, each component of which is immiscible with the others (middle vial), is emulsified (left vial) using surfactant FC6–HC6. Right vial shows a two-component system, water/perfluoropolyether oil, which is emulsified using the same surfactant.

7.3 Anomalous micelle behavior in dilute solutions

Many studies on mixed systems of fluorocarbon and hydrocarbon surfactants have revealed that extremely heterogeneous micelles are formed in aqueous solution [11–15]. Such heterogeneity may also be present among micelles of the hybrid surfactants, which have both hydrocarbon and fluorocarbon chains in their molecules. It is therefore worthwhile studying the aggregation state of the hybrid surfactant micelle as a model for the fluorocarbon–hydrocarbon mixed micelle. The aggregation state of fluorocarbon–hydrocarbon hybrid surfactant micelles has been studied by Guo *et al.* [6] and Inoue *et al.* [16] by means of ^{1}H and ^{19}F NMR. Guo *et al.* [6] reported that both fluorocarbon and hydrocarbon chains are incorporated in the interior of the micelle when the hydrocarbon chain bears three carbon atoms or more. Inoue *et al.* [16] studied the surfactants with a short fluorocarbon chain and a long hydrocarbon chain, $C_nF_{2n+1}CONHCH_2CH_2N^+(CH_3)_2C_{16}H_{33}Br^-$ ($n = 1, 2, 3$) and showed that fluoroalkyl groups extend their chains straight forward to the micellar surface and that the terminal CF_3 groups directly face the bulk phase water. Of course, this results from the synthesized control, where the alkyl groups are long enough to singly form the hydrophobic core and expel the short fluorocarbon chain to locate at the surface. In this section, the internal micelle environment in aqueous solutions of the hybrid surfactants bearing different hydrocarbon and fluorocarbon chains is examined by fluorescence probe and Raman spectroscopic methods [17, 18].

Figure 7.4 shows the relationship between surface tension and the concentration of FC6–HC4 in aqueous solution. The surface tension decreases with an increase in surfactant concentration and the curve inflects to give the CMC, which is equal to the value obtained from static light-scattering measurements. Similar behavior is observed with the other hybrid surfactants, except CF6–HC6. It should be noted that the surface tension continues to decrease slightly even above the CMC. Such a lingering decrease of surface tension above the CMC has been reported for mixed micelle solutions of fluorocarbon and hydrocarbon surfactants, and some single surfactants such as sodium dodecyl sulfate (SDS) and Aerosol OT (AOT, sodium 2-ethylhexyl sulfosuccinate) [19]. In the mixed micelle systems, the secondary decrease is interpreted as indicating the coexistence of two different types of micelles: fluorocarbon surfactant-rich micelles and hydrocarbon surfactant-rich micelles. In the case of single surfactants such as FC*m*–HC*n*s, however, two species of homogeneous micelles are not thought to coexist because there is a fixed ratio of fluorocarbon chain length to hydrocarbon chain length. Nevertheless, this probably does occur because the micellar structure changes with a concentration increment above the CMC, as described later.

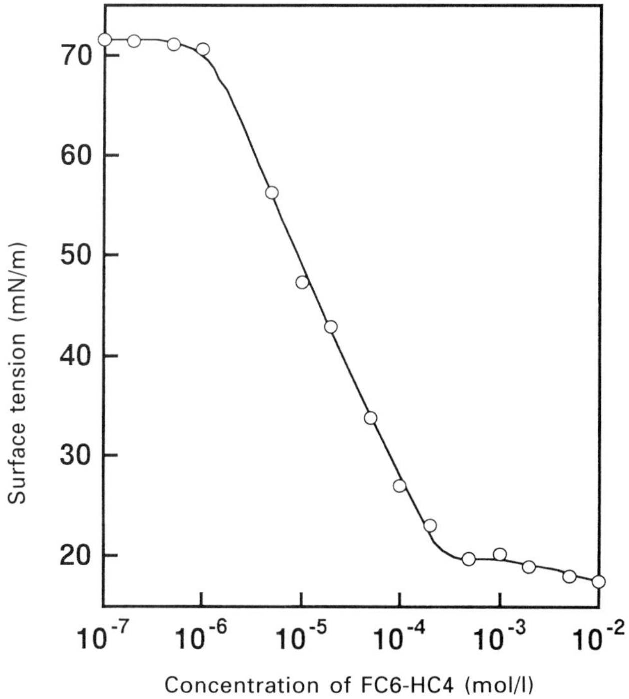

Figure 7.4 Surface tension against concentration for FC6–HC4 at 25°C.

In order to evaluate the change in the micelle aggregating condition of the hybrid surfactant according to its concentration increment, the internal polarity of the micelle was monitored by pyrene fluorescence. The fluorescence spectrum of pyrene has five predominant vibronic peaks, numbered 1 to 5 from the shorter to the longer wavelength. The intensity ratio of the first peak to the third one, I_1/I_3, indicates the micropolarity around pyrene molecules at the solubilizing site [20, 21].

Figure 7.5 shows the concentration dependence of the I_1/I_3 ratio of SDS, a typical anionic hydrocarbon surfactant, and that of a mixture of SDS and sodium perfluorooctanoate (SPFO), an anionic fluorocarbon surfactant, at a 1:1 molar ratio. In both solutions the I_1/I_3 ratio below the CMC is about 1.56, which is equal to the value in water. The ratio decreases rapidly to about 1.02 at the CMCs. Since such a low value is characteristic for pyrene in hydrocarbons, the decrease indicates the formation of micelles and the transfer of pyrene molecules from bulk water to the micelles. However, the concentration dependence of the I_1/I_3 ratio of FC6–HC4 solutions is completely different from that of the above-mentioned surfactants, as shown in Figure 7.6.

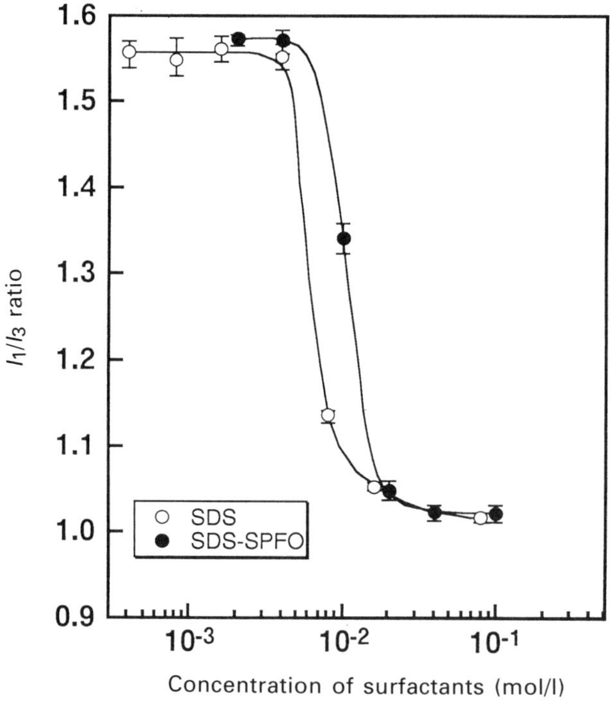

Figure 7.5 Relationship between I_1/I_3 ratio and concentration for SDS and SDS/SPFO at 25°C.

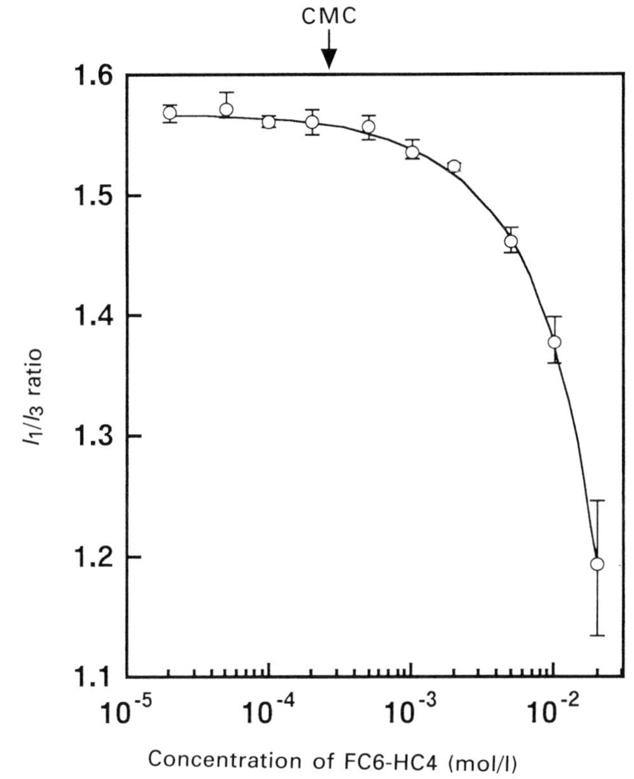

Figure 7.6 Concentration dependence of I_1/I_3 for FC6–HC4 at 25°C.

A behavior similar to that in Figure 7.6 is observed with the other hybrid surfactants. The I_1/I_3 ratio for FC6–HC4 solutions below the CMC is almost the same as that for the SDS and SDS–SPFO solutions shown in Figure 7.5, and the slight decrease in I_1/I_3 value could be ascribed to experimental errors, judging from the width of the error bar. However, unlike the SDS and SDS–SPFO mixture systems, the value decreases gradually above the CMC and then decreases further down to about 1.20 when the concentration becomes one hundred times higher than the CMC. This suggests that the micelle of the hybrid surfactant gradually changes its aggregation state, e.g. the orientation of the hydrocarbon moiety and/or packing density of the surfactant, above the CMC as the concentration increases. In highly concentrated solutions, pyrene molecules are likely to be solubilized within the micelles as they are in SDS and SDS–SPFO solutions. Correspondingly, the maximum absorbance wavelength in the ultraviolet spectrum of pyrene is almost invariant at 334 nm, independent of FC6-HC4 concentration, and is close to the 333.5 nm observed in *n*-hexane, a model of hydrocarbon environment, rather than the 327.1 nm observed in perfluoro-*n*-hexane, a model of fluorocarbon environment [18].

Pyrene molecules are thought to be solubilized in the hydrocarbon moiety of FC*m*–HC*n* micelles, but the environment around pyrene changes gradually from one of large micropolarity to one of small micropolarity with an increase in the concentration of FC*m*–HC*n*s. As described previously, FC*m*–HC*n*s cannot form two species of homogeneous micelles with different fluorocarbon–hydrocarbon composition because of the fixed ratio of fluorocarbon chain to hydrocarbon chain. Hence, the anomalous I_1/I_3 behavior is ascribable to the change in the aggregation state of FC*m*–HC*n* micelles.

If the micellar structures of FC*m*–HC*n*s change according to their concentration increments, they may have a second CMC. If they form one type of micelle at the first CMC (c_1) and another type of micelle at the second CMC (c_2), the I_1/I_3 value observed consists of the I_1/I_3 ratio in bulk $(I_1/I_3)_0$, the ratio in the first micelle $(I_1/I_3)_1$ and the ratio in the second micelle $(I_1/I_3)_2$. Below the first CMC, i.e. when the total surfactant concentration, C, is less than c_1, the I_1/I_3 value observed is expressed by equation (7.2):

$$\frac{I_1}{I_3} = \left(\frac{I_1}{I_3}\right)_0 \tag{7.2}$$

Moreover, the ratio is given by equation (7.3) in the concentration range between c_1 and c_2, i.e. when $c_1 < C < c_2$:

$$\frac{I_1}{I_3} = \frac{\left(\frac{I_1}{I_3}\right)_0 c_1 + \left(\frac{I_1}{I_3}\right)_1 (C - c_1)}{C} = c_1\left[\left(\frac{I_1}{I_3}\right)_0 - \left(\frac{I_1}{I_3}\right)_1\right]\frac{1}{C} + \left(\frac{I_1}{I_3}\right)_1 \tag{7.3}$$

and by equation (7.4) above c_2, i.e. when $c_2 < C$:

$$\left(\frac{I_1}{I_3}\right) = \frac{\left(\frac{I_1}{I_3}\right)_0 c_1 + \left(\frac{I_1}{I_3}\right)_1 (c_2 - c_1) + \left(\frac{I_1}{I_3}\right)_2 (C - c_2)}{C}$$

$$= \left\{c_1\left[\left(\frac{I_1}{I_3}\right)_0 - \left(\frac{I_1}{I_3}\right)_1\right] + c_2\left[\left(\frac{I_1}{I_3}\right)_1 - \left(\frac{I_1}{I_3}\right)_2\right]\right\}\frac{1}{C} + \left(\frac{I_1}{I_3}\right)_2 \tag{7.4}$$

Plots of the experimental data according to these equations give a straight line when the I_1/I_3 ratio and the reciprocal of surfactant concentration are taken as the ordinate and abscissa, respectively, and the slope of the straight line furnishes and changes in the micelle condition.

Figure 7.7 shows the relationship between the I_1/I_3 ratio and the reciprocal concentration for FC4–HC4 and FC6–HC4. In the case of FC4–HC4 (Figure 7.7a), the concentration (c_1) corresponding to the intersection of the two lines is almost identical to the CMC obtained by the surface tension measurement, while the second inflection is neither clear enough nor close to c_1. FC6–HC4 (Figure 7.7b), however, shows two intersections, one at c_1 and the other at c_2. This implies that the surfactant has the second CMC in addition to the first one. The highly critical behavior of these surfactants is found to be ascribable to the length of hydrocarbon or fluorocarbon chain. The $(I_1/I_3)_0$ values for FC*m*–HC*n*s are similar to the value of I_1/I_3 for pyrene dissolved in water. The values of $(I_1/I_3)_1$ for FC4–HC6 and the FC6–HC*n* series are 1.46–1.55, indicating that the micropolarity of hydrocarbon chains is high in the first type of micelles, i.e. water molecules penetrate deeply into the micelles or the hydrocarbon chains involved in the micelles exist near the bulk phase. In contrast, the values of $(I_1/I_3)_1$ and $(I_1/I_3)_2$ are as low as the value of I_1/I_3 for pyrene in SDS micellar solutions, which means that the hydrocarbon chains locate at the hydrophobic site in the first or second type of micelles. Correspondingly, the presence of two intersections reflects the formation of wet, preliminary micelles prior to ordinary micelle formation, with much poorer adsorption of pyrene on a single kind of micelle, which is seen for SPFO.

Raman spectroscopy has been used to investigate the change in the microenvironment of the micellar interior in the high concentration range above the CMC because the Raman intensity for FC*m*–HC*n* below their CMCs is too weak to measure. In fact, the intensity of the CH_3 symmetric stretching vibration relative to its asymmetric stretching vibration increases with increasing concentration only for the FC6–HC*n* series [18], which implies that the degree of disorder of hydrocarbon chains increases with increasing concentration of surfactant [22, 23].

Figure 7.8 shows the concentration dependence of the wavenumber of the CH_3 asymmetric stretching vibration. In the case of FC4–HC*n*s the wavenumber is almost constant, independent of the increase in the surfactant concentration, whereas in the case of FC6–HC*n*s, the wavenumber shifts downwards as the concentration increases and becomes invariant around the second CMC. A similar trend is also found for the CH_3 symmetric stretching bands. This behavior clearly demonstrates the presence of the second CMC because it is free from the probe (like pyrene) that relates to the adsorption isotherm. Kamogawa *et al.* have reported that dehydration of C–H results in a reduction of the wavenumber [24]. In other words, dehydration of alkyl chains involved in the surfactants may take place at a high concentration of FC6–HC*n* solutions.

(a)

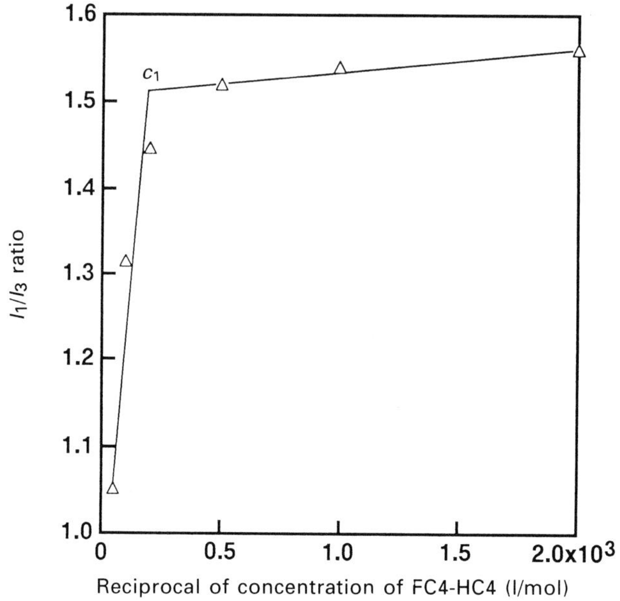

(b)

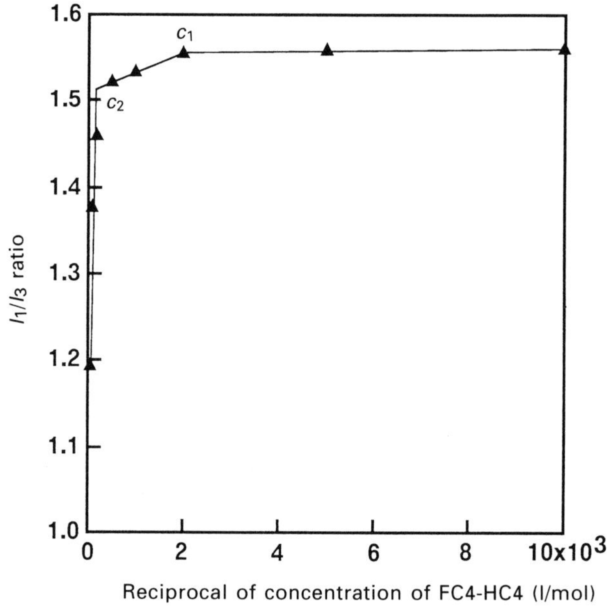

Figure 7.7 Relationship between I_1/I_3 and reciprocal concentration for (a) FC4–HC4 and (b) FC6–HC4 at 25°C.

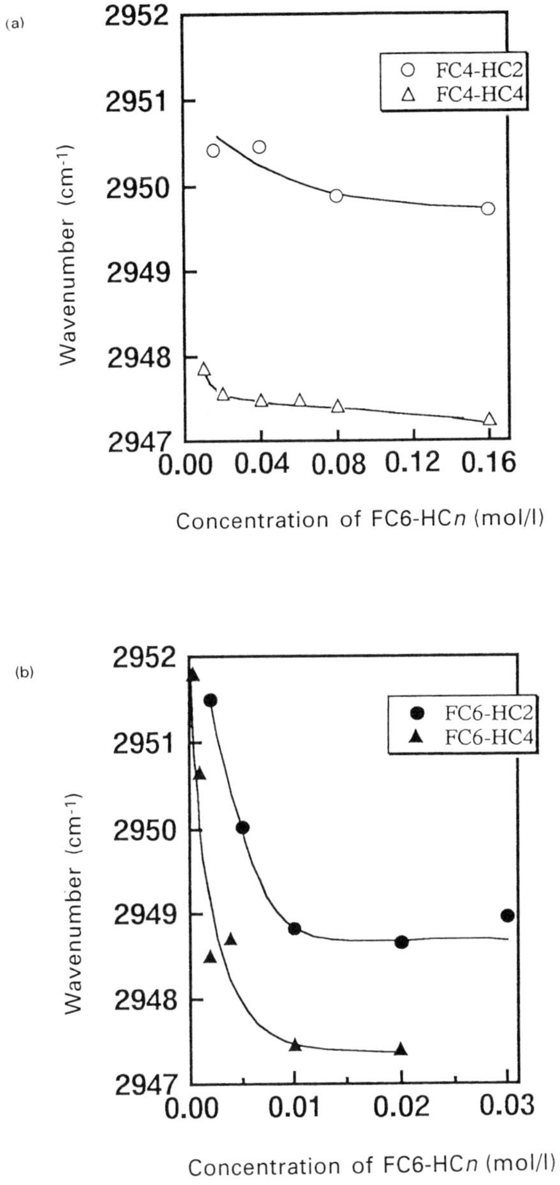

Figure 7.8 Concentration dependence of Raman wavenumber of CH_3 asymmetric stretching band for (a) FC4–HC*n* and (b) FC6–HC*n* at 25°C.

The wavenumbers of the C–C stretching vibration of the phenyl ring observed at 980–1000 cm^{-1} and the SO_3 symmetric stretching vibration observed at about 1050 cm^{-1} decrease with an increase in the concentration of the surfactants [18]. The phenyl ring of FC*m*–HC*n*s is supposed to be more hydrated at higher surfactant concentrations. Furthermore, the wavenumber of the SO_3 symmetric stretching vibration increases with an increase in the surfactant concentration [18], suggesting an increasing interaction between the SO_3 and the Na ion with increasing surfactant concentration [25].

In the case of FC6–HC*n*s and FC4–HC6, for which the second CMC is observed, water molecules may invade the interior of micelles above the CMC (micelle 1) to give them a loose structure, resulting in high I_1/I_3 values. At concentrations above the second CMC (micelle 2), water molecules are removed from the micelle interior, giving low I_1/I_3 values. On the other hand, the first type of micelle is rather unstable and the second type of micelle dominates in solutions of the other FC4–HC*n*s. The instability is thought to be caused by the short chain length of FC or HC. Consequently, the model shown in Figure 7.9 has been adopted. In the FC4–HC*n* series micelle 1 is always formed, while in the FC6–HC*n* series the micellar structure is altered from that of micelle 1 to that of micelle 2 with increasing concentration.

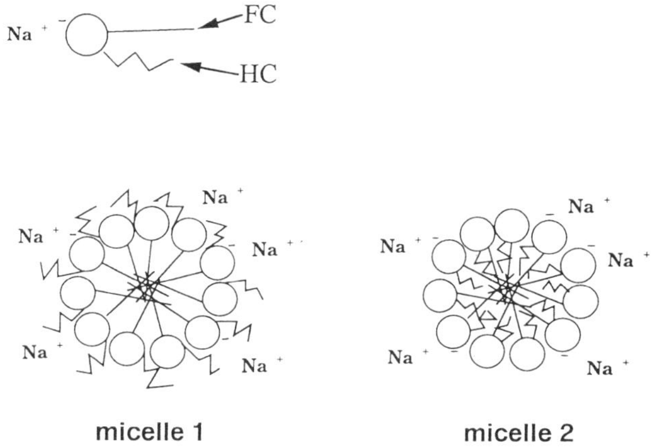

Figure 7.9 Illustration of micelle 1 and micelle 2.

7.4 Composition-dependent viscoelasticity of concentrated solutions

The purpose of this section is to discuss the anomalous viscoelastic behavior of concentrated solutions of the hybrid surfactants.

Figure 7.10 shows the change observed at the solution surface when containers with 1, 10, 20 and 30 wt% solutions of FC6–HC4 are tilted from the upright position [26]. It can be seen in that the surface of all the solutions, except that of the 10 wt% solution, inclines immediately after the containers are tilted. The surface of the 10 wt% solution inclines approximately 5 s later. It should be noted that the solution diluted to 10 wt% from 20 or 30 wt% solutions also shows this behavior, indicating the higher viscosity of the 10 wt% solution. The 10 wt% solution has a characteristic rubber-like elasticity when a glass rod is pulled from it (Figure 7.11), which the 1, 20 and 30 wt% solutions do not possess.

Figure 7.10 Change in the surfaces of aqueous 1, 10, 20 and 30 wt% solutions of FC6–HC4 when the containers are tilted from an upright position at room temperature.

Figure 7.11 Rubber-like behaviour of 10 wt% solution of FC6–HC4 when a glass rod is pulled from the solution at room temperature.

The flow properties of the 1, 10, 20 and 30 wt% solutions have also been analyzed. The viscosity, defined as the slope (shear stress/shear rate) around the zero shear rate, is largest for the 10 wt% solution (Figure 7.12). The flow properties and dynamic viscoelasticity of the surfactant solutions were measured using a stress-controlled rheometer (Carri-Med CSL^{2}100, TA-instruments Co.). A double concentric cylinder system and a cone plate sensor with a 2° cone angle and 6 cm in diameter were used to measure flow properties and dynamic viscoelasticity, respectively. Shear stress is continuously altered in the range of 0 to 50 Pa at 25±0.1 °C while measuring the flow properties. The dynamic viscoelasticity was analyzed within a range of strain that gave a linear relationship with the complex modulus for each sample. The angular frequency was 1 to 100 rad/s during the analysis. Figure 7.12 also indicates the following:

- the 1 wt% solution has rheological characteristics similar to those for Newtonian fluids in that the viscosity is not dependent on the shear rate and does not show any sign of hysteresis
- the flow curve of the 10 wt% solution shows inconsistency at a high shear rate, and also shows hysteresis
- the rheological property of the 20 wt% solution is similar to that of Newtonian fluids in that the viscosity is not dependent on the shear rate, and does not show any sign of hysteresis
- the viscosity of the 30 wt% solution decreases proportionally to the shear rate increase, but does not show hysteresis.

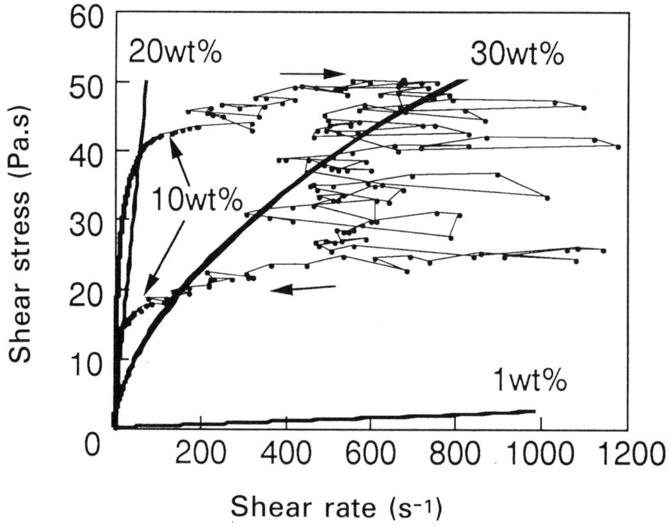

Figure 7.12 Flow curves for aqueous 1, 10, 20 and 30 wt% solutions of FC6–HC4 at 25°C.

The relationship between viscosity and shear rate is shown in Figure 7.13. The viscosity maintains a constant value of approximately 12 Pa.s until the shear rate reaches a value of *c.* 20 s^{-1}, then it rapidly decreases to around 10^{-2} Pa.s (curve a). When the shear rate is reduced, hysteresis appears (curve b). If another measurement is performed on the sample to which the same magnitude of shear rate was applied before, it follows the viscosity change shown in curve b. Thus, the viscosity of the 10 wt% solution irreversibly changes when a shear stress is applied. When a shear stress small enough not to decrease the viscosity is applied to the solution, the viscosity returns to the original curve (the flat part of curve a).

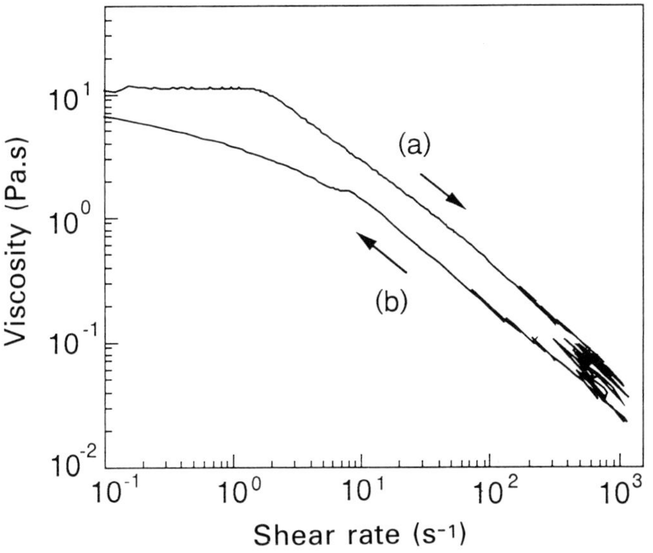

Figure 7.13 Correlation between viscosity and shear rate for aqueous 10 wt% solution of FC6–HC4 at 25°C.

Generally, the viscosity of single-chain hydrocarbon or single-chain fluorocarbon surfactant tends to increase uniformly with an increase in the surfactant concentration. However, it is a unique characteristic of the hybrid surfactant that the viscosity reaches a maximum value at a certain concentration.

Figure 7.14 shows the dependency of angle frequency on storage modulus (G') for aqueous solutions of FC6–HC4. No measurement was made on the 1 wt% solution because it had practically no elasticity. Generally, the storage modulus of concentrated suspensions shows no angular frequency dependency and this is assumed to be caused by the relaxation of a temporarily formed internal structure such as a network consisting of dispersed molecules [27]. The 10 wt% solution shows a significant angular frequency dependency in low and high frequency regions, and shows a high storage modulus in a high frequency region. This suggests that the 10 wt% solution also has a structure that is different from that observed with the 30 wt% solution [28]. The storage modulus of the 20 wt% solution has a certain degree of angular frequency dependency but the values are relatively low, suggesting formation of few or no aggregates in the 20 wt% solution. Although G' in 30 wt% solution shows no angular frequency dependency, the storage modulus is relatively high, implying that the same phenomenon occurs in the 30 wt% solution.

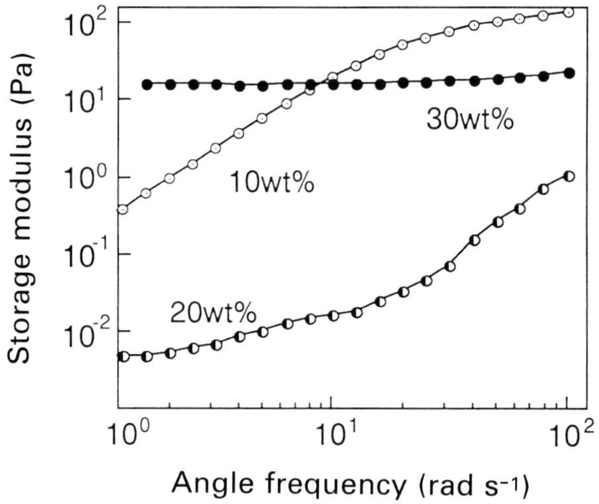

Figure 7.14 Dependence of storage modulus on angle frequency for aqueous solutions of FC6–HC4 at 25°C.

Figure 7.15 shows the correlation between the loss tangent (tan δ = loss modulus/storage modulus) and the concentration of FC6–HC4 solution. The loss tangent, tan δ, becomes a minimum around 10 wt%, indicating that the 10 wt% solution has not only elasticity but also viscosity. While the 30 wt% solution also has a low tan δ value, its G' value is smaller than that of the 10 wt% solution. Thus, the 30 wt% solution has an elasticity lower than that of the 10 wt% solution because a lamellar liquid crystalline phase is formed in the 30 wt% solution.

In order to investigate the reason why this hybrid surfactant (FC6–HC4) shows a viscoelasticity when its concentration is around 10 wt%, changes in the structure of water molecules and in the aggregating condition of the surfactant were considered. Attempts have been made by NMR to detect the structural change of water molecules as the concentration of the surfactant increases. The self-diffusion coefficient (D) of water is obtained by the pulsed gradient spin-echo NMR technique and the ^{1}H NMR spin-lattice relaxation time (T_1) of water is measured by the inversion-recovery method using a 400 Mhz NMR spectrometer (JEOL EX-400) at 30°C. However, both D and T_1 decrease uniformly with increasing surfactant concentration, demonstrating that the motion of water molecules is not restricted in solutions at around 10 wt%.

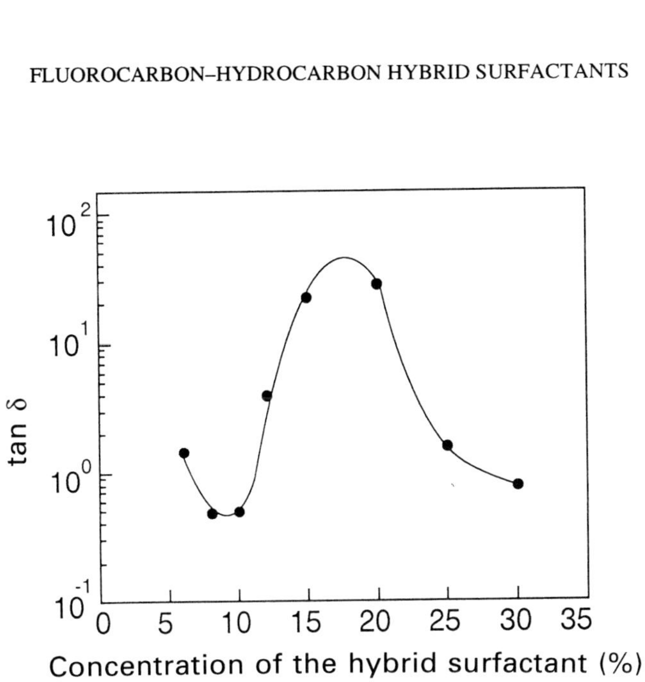

Figure 7.15 Relationship between tan δ (loss modulus/storage modulus) and the concentration of FC6–HC4 at 100.0 rad/s angle frequency (25°C).

As regards the reduction of viscosity with increasing solute concentration, Baek *et al.* [29] reported that the formation of liquid-crystalline hydroxypropylcellulose (a water-soluble polymer) results in a reduction in viscosity. In fact, the hybrid surfactant (FC6–HC4) used here does form a liquid-crystalline hydroxypropylcellulose at concentrations of 20 wt% or higher but not at 10 wt%. However, for this hybrid surfactant, the internal structure of the micelle varies depending on the increase of concentration in the diluted condition [18, 19]. It is therefore assumed that the abnormal viscoelasticity in surfactant solutions of high concentrations is mainly caused by a surface orientation change resulting from the structural change in the micelle brought about by the increase in concentration.

7.5 Thermoresponsive viscoelasticity of concentrated solutions

This section describes the temperature-dependence of the viscoelasticity of aqueous FC6–HC4 solution.

Viscosity measurement is conducted in such a way that shear stress is increased from 0 to 1.0 Pa within 30 s and the shear stress versus shear rate curve obtained is approximated by the Bingham equation. Dynamic viscoelasticity measurement is performed in the linear range of percentage strain after finding a linear range for each sample. The parameters for dynamic viscoelasticity are determined at a fixed angular frequency of 1.0 Hz.

The relationship between viscosity and concentration for aqueous FC6–HC4 solution at 25°C is shown in Figure 7.16. A maximum appears in the viscosity at a surfactant concentration of about 10 wt%, as mentioned previously. Although the viscosity lowering for the polymer solution at high concentrations is ascribed to the formation of liquid crystals, this explanation cannot be true for aqueous FC6–HC4 solution because no liquid crystal formation is observed at concentrations around 10 wt%. Liquid crystal formation for this surfactant is found, using a polarization microscope, to start at the much higher concentration of 30 wt%.

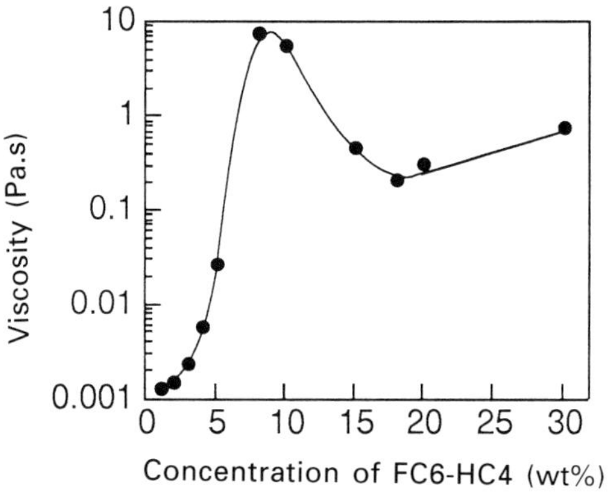

Figure 7.16 Relationship between viscosity and concentration for FC6–HC4 solution at 25°C.

The anomalous viscoelasticity behavior of aqueous FC6–HC4 solution observed when the temperature is raised is shown in Figure 7.17. The viscosity reduces with rising temperature at surfactant concentrations of 20 and 30 wt%, while it increases from about 0.9 Pa.s to about 90 Pa.s (10^2 times) as the temperature rises from 6 to 36°C (showing a maximum) and then decreases to about 0.009 Pa.s at 64°C (10^{-4} times) with further temperature rise when the surfactant concentration is 10 wt%. Such temperature-dependent viscosity change is referred to as thermoresponsive viscoelasticity. Thermoresponsive viscoelasticity is observed with this surfactant at least in the concentration range 7 to 15 wt% and the value of the maximum viscosity and the temperature at which maximum viscosity appears are dependent on the surfactant concentration. Visual observations reveal that the temperatures at which a minimum appears on the curve in Figure 7.17, 6°C for 10 wt% solution and 22°C for 20 wt% solution, correspond to the temperatures at which the surfactant deposits. No surfactant deposition is found in the 30 wt% solution in the temperature range used in this work because the surfactant forms lamellar liquid crystals at this concentration. Moreover, thermoresponsive viscoelasticity is not observed with aqueous solutions of monohydrocarbon-chain-type surfactants (such as 1-pentanesulfonic acid sodium salt and sodium dodecyl sulfate), monofluorocarbon-chain-type surfactants (such as perfluoroheptanoate), and the other fluoro-hybrid-type surfactants such as FC4–HC6. In other words, thermoresponsive viscoelastic behavior is specific for FC6–HC4 and observed only with its aqueous solutions in a certain concentration range.

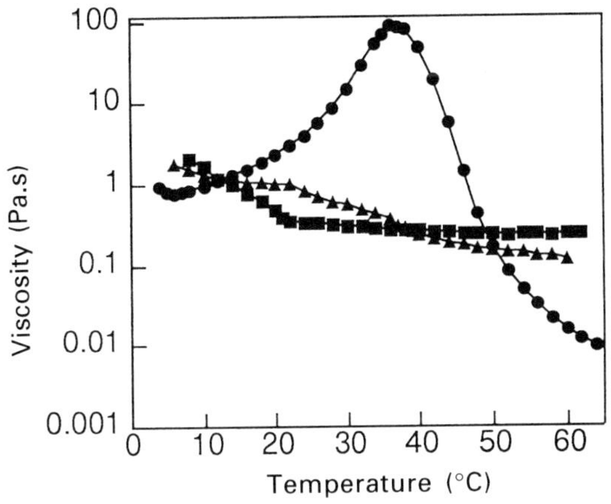

Figure 7.17 Temperature dependence of viscosity for aqueous FC6–HC4 solutions:

Dynamic viscoelasticity measurement has been performed exclusively on an aqueous 10 wt% FC6–HC4 solution that exhibited thermoresponsive viscoelastic behavior at a fixed angular frequency (1.0 Hz) while the temperature of the solution is varied. Although both storage modulus (G') and loss modulus (G'') change, showing an upward convex curve with increasing temperature, they do not necessarily exhibit a peak at 36°C (Figure 7.18). In other words, the temperature of 36°C is strictly the temperature at which the ratio of elastic component and viscous component or tan δ (= G''/G') becomes a minimum. The spin-lattice relaxation time, T_1, for protons increases with rising temperature and decreases with increasing surfactant concentration, both in a uniform way. Similarly, the self-diffusion coefficient of water, D, increases uniformly with rising temperature and decreases uniformly with increasing surfactant concentration. No unusual behavior in these parameters is found either at 10wt% FC6–HC4 concentration or at 36°C. This suggests that the hydration structure of water molecules does not contribute much to the viscoelastic body formation in solutions of the hybrid surfactant, judging from the NMR and ultramicro-regional observations.

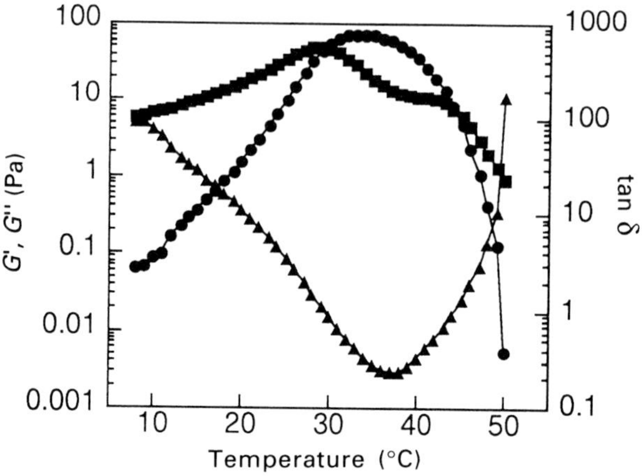

Figure 7.18 Temperature dependence of *G'*, *G''* and tan δ for aqueous FC6–HC4 (frequency = 1.0 Hz).

Figure 7.19 shows the relationship between viscosity and shear rate for 10 wt% FC6–HC4 aqueous solution at different temperatures [30]. In the low shear rate region, the viscosity is considerably dependent on temperature, while in the high shear rate region it begins to decrease rapidly and attains almost the same value independently of temperature. This suggests that the loosely assembled structure formed by surfactant molecules in the solution must break down because the fluidity of surfactant molecules after destruction of the structure is almost independent of temperature, thus resulting in the same viscosity behavior of the solution. Therefore, the temperature-dependent solution viscosity in the low shear rate region suggests that the shape or degree of interwoven colloidal assemblies is also temperature dependent. As can be seen in Figure 7.19, the viscosity at 36°C is the highest, which implies temporary formation of a rather rigid internal structure, such as a network consisting of rod-like micelles, at this temperature.

Further studies are in progress and the results will be published in the near future.

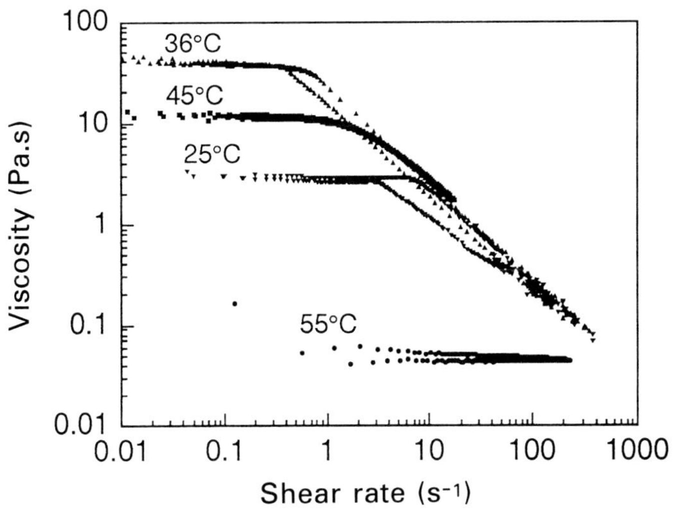

Figure 7.19 Correlation between viscosity and shear rate for aqueous 10 wt% FC6–HC4 solution at different temperatures.

References

[1] Yoshino, N., Komine, N., Suzuki, J., Arima, Y. and Hirai, H. (1991) Syntheses of anionic surfactants having two polyfluoroalkyl chains and their flocculation ability for dispersed magnetite particles in water. *Bulletin of the Chemical Society of Japan*, **64** 3262.

[2] Shinoda, K., Hato, M. and Hayashi, T. (1972) Physicochemical properties of aqueous solutions of fluorinated surfactants. *Journal of Physical Chemistry*, **76** 909.

[3] Abe, M., Yamaguchi, T., Shibata, Y., Uchiyama, H., Yoshino, N., Ogino, K. and Christian, S.D. (1992) Solution properties of mixed fluorocarbon–hydrocarbon surfactant systems. *Colloids Surf.*, **67** 29.

[4] Burkitt, S. J., Ottewill, R. H., Hayter, J. B. and Ingram, B. T. (1987) Small-angle neutron scattering studies on micellar systems. Part 2. Mixed systems of ammonium decanoate and ammonium perfluorooctanoate. *Colloid Polymer Science*, **265** 628.

[5] Funasaki, N. and Hada, S. (1983) Quaternary structure and spin-state transition in azide methemoglobin A. *Journal of Physical Chemistry*, **87** 342.

[6] Guo, W., Li, Z., Fung, B.M., O'Rear, E.A. and Harwell, J.H. (1992) Hybrid surfactants containing separate hydrocarbon and fluorocarbon chains. *Journal of Physical Chemistry*, **96** 6738.

[7] Guo, W., Fung, B.M. and O'Rear, E.A. (1992) Exchange of hybrid surfactant molecules between monomers and micelles. *Journal of Physical Chemistry*, **96** 10068.

[8] Yoshino, N., Hamano, K., Omiya, T., Kondo, Y., Ito, A. and Abe, M. (1995) Synthesis of hybrid anionic surfactants containing fluorocarbon and hydrocarbon chains. *Langmuir*, **11** 466.

[9] Ito, A., Sakai, H., Kondo, Y., Yoshino, N. and Abe, M. (1996) Micellar solution properties of fluorocarbon–hydrocarbon hybrid surfactant. *Langmuir*, **12** 5768.

[10] Hato, M. (1977) Physical chemistry of surfactant solutions. *Hyomen*, **15** 372.

[11] Mukerjee, P. and Yang, A.Y.S. (1976) Nonideality of mixing of micelles of fluorocarbon and hydrocarbon surfactants and evidence of partial miscibility from differential conductance data. *Journal of Physical Chemistry*, **80** 1388.

[12] Carlfors, J. and Stilbs, P. (1984) The composition of mixed micelles of fluorocarbon and hydrocarbon surfactants as derived from nuclear magnetic resonance self-diffusion measurements. *Journal of Physical Chemistry*, **88** 4410.

[13] Muto, Y., Esumi, K., Meguro, K. and Zana, R. (1987) Aggregation behavior of mixed fluorocarbon and hydrocarbon surfactants in aqueous solutions. *Journal of Colloid Interface Science*, **120** 162.

[14] Kalyanasundaram, K. (1988) Pyrene fluorescence as a probe of fluorocarbon micelles and their mixed micelles with hydrocarbon surfactants. *Langmuir*, **4** 942.

[15] Kamogawa, K. and Tajima, K. (1993) Distinctive behavior of the ESR correlation times of DOXYL spin probes in mixed micellar solutions of hydrocarbon and fluorocarbon surfactants. *Journal of Physical Chemistry*, **97** 9506.

[16] Inoue, H., Arai, S., Kakuta, Y., Taki, M., Masuda, H., Moronuki, N. and Yamada, M. (1992) *Mem. Fac. Technol., Tokyo Metrop. Univ.*, **42** 4511.

[17] Ito, A., Komogawa, K., Sakai, H., Hamano, K., Kondo, Y., Yoshino, N. and Abe, M. (1996) Anomalous behavior of fluorocarbon–hydrocarbon hybrid surfactant in its aqueous solution. Surface tension and micropolarity. *Journal of the Japan Oil Chemists' Society*, **45** 479.

[18] Ito, A., Komogawa, K., Sakai, H., Hamano, K., Kondo, Y., Yoshino, N. and Abe, M. (1997) Micelle aggregating condition of fluorocarbon–hydrocarbon hybrid surfactants in aqueous solution. *Langmuir*, **13** 2935.

[19] Purcell, I.P., Thomas, R.K., Penfold, J. and Howe, A.M. (1995) The composition and structure of sodium dodecyl sulfate–decanol mixtures adsorbed at the air–water interface: a neutron reflection study. *Colloids Surf.*, **94** 125.

[20] Kalyanasundaram, K. and Thomas, J.K. (1977) Environmental effects on vibronic band intensities in pyrene monomer fluorescence and their application in studies of micellar systems. *Journal of the American Chemical Society*, **99** 2039.

[21] Turro, N.J., Kuo, P.L., Somasundaran, P. and Wong, K. (1986) Photoluminescence probes for pressure and temperature effects on the aggregates of water-soluble block copolymers. *Journal of Physical Chemistry*, **90** 288.

[22] Amorim da costa, A.M., Geraldes, C.F.G.C. and Teixeira-dias, J.C. (1982) Micellar aggregation of CTAB in water and chloroform solutions—a study by laser Raman spectroscopy. *Jornal of Colloid Interface Science*, **86** 254.

[23] Maitra, A. and Jain, T.K. (1987) Study of the vibrational characteristics of aerosol OT by laser Raman spectroscopy. *Colloids Surf.*, **28** 19.

[24] Kamogawa, K., Tajima, K., Hayakawa, K. and Kitagawa, T. (1984) Raman spectroscopic studies of submillimolar surfactant solutions. Concentration dependence of the carbon-hydrogen stretching Raman lines. *Journal of Physical Chemistry*, **88** 2494.

[25] Kawai, T., Umemura, J. and Takenaka, T. (1984) Fourier transform infrared spectra and micelle formation of sodium *n*-alkanesulfonates in aqueous solution. *Colloid Polym. Sci.*, **262** 61.

[26] Abe, M., Tobita, K, Sakai, H., Kondo, Y., Yoshino, N., Kasahara, Y., Matsuzawa, H., Iwahashi, M., Momozawa, N. and Nishiyama, K. (1997) Anomalous viscoelasticity of concentrated solutions with a fluorinated hybrid surfactant. *Langmuir*, **13** 2932.

[27] Amari, T. (1986) Rheological properties of disperse systems of pigments. *Nihon Rheology Gakkaishi*, **14** 167-76.

[28] Kaneko, T., Hayashi, T. and Amari, T. (1995) Rheological study on transfer and splitting of carbon black/alkyd resin suspensions at roll nip. *Journal of the Japan Society of Colour Material*, **68** 673.

[29] Baek, S.G., Magda, J.J., Larson, R.G. and Hudson, S.D. (1994) Rheological differences among liquid-crystalline polymers. II. Disappearance of negative N_1 in densely packed lyotropes and thermotropes. *Journal of Rheology*, **38** 1473.

[30] Tobita, K., Sakai, H., Kondo, Y., Yoshino, N., Iwasawa, M., Momozawa, N. and Abe, M. (1997) Thermoresponsive viscoelasticity of sodium 1-oxo-1-[4-(tridecafluorohexyl]-2-hexanesulfonate aqueous solutions. *Langmuir*, **13** 5054.

8 Polymeric surfactants

Th. F. Tadros

8.1 Background

Polymeric surfactants of the A–B, A–B–A block and BA_n graft type are ideal molecules for stabilisation of suspensions and emulsions. The B chain is chosen to have very high affinity (strong adsorption) for the surface of particles or droplets, whereas the A chains (the stabilising chains) are chosen to be highly solvated by the continuous medium. This chapter describes the principles involved in the stabilisation of suspensions and emulsions by polymeric surfactants. The first part gives a general classification of polymeric surfactants, followed by a section on their solution properties. The second part describes the adsorption and conformation of polymeric surfactants at the interface. The methods that can be applied to determine the adsorption and conformation of polymeric surfactants are briefly described. The third part deals with the stabilisation mechanism produced using polymeric surfactants. The main repulsive forces are considered, namely mixing or osmotic repulsion and the elastic, volume restriction or entropic repulsion. Combination of the mixing and elastic contributions with the van der Waals' attraction results in an energy–distance curve that shows weak attraction at a separation distance, h, comparable to twice the adsorbed layer thickness, 2δ. However, when $h < 2\delta$, very strong repulsion arises when h is further reduced. The main criteria for effective steric stabilisation are summarised. These are: complete coverage of the particle surface by the polymeric surfactant, strong adsorption (anchoring) of the chains to the surface, good solvency of the chains by the continuous medium and reasonably thick adsorbed layers (5–10 nm). These criteria are best satisfied by block and graft copolymers. The fourth part of the chapter gives an illustration of the use of graft copolymers for stabilisation of suspensions. The interparticle interaction in these systems can be studied using rheological measurements. Using graft copolymers, highly concentrated suspensions with long term colloid stability can be prepared. The next part deals with stabilisation of oil-in-water (o/w) and water-in-oil (w/o) emulsions using block copolymers. Very stable concentrated emulsions can be prepared using block copolymers. These emulsions can be investigated using rheological measurements. The final part of the chapter deals with the preparation of stable water-in-oil-in-water (w/o/w) multiple emulsions that are suitable for application in crop protection or personal-care formulations. The main criterion for producing stable w/o/w systems is the use of two polymeric surfactants, one with a low hydrophilic–lipophilic balance (HLB) number for preparation of the primary w/o emulsion and a second one with a high HLB number for preparation of the final w/o/w multiple emulsion. The primary emulsifier should produce a viscoelastic film that will prevent leakage from the internal water droplets to the outside continuous phase. The secondary emulsifier should also provide an effective barrier against flocculation and coalescence of the multiple emulsion on storage.

8.2 Introduction

Surfactants (amphipathic molecules consisting of hydrophobic and hydrophilic portions) are essential materials for the preparation of many industrial systems, for example dyestuffs, paper coatings, inks, agrochemicals, pharmaceuticals, personal-care products, ceramics and detergents. One of the most important applications of surfactants is in the preparation of o/w and w/o emulsions. In these emulsions the hydrophobic portion of the surfactant molecule should adsorb strongly at the o/w interface or dissolve in the oil phase, leaving the hydrophilic components in the aqueous medium where they become strongly solvated by water molecules. This configuration provides an effective repulsive force, usually referred to as steric stabilisation. A convenient method for selection of surfactants is based on the relative proportions of the hydrophilic and hydrophobic groups, i.e. the hydrophilic–lipophilic balance (HLB) introduced by Griffin [1, 2]. For o/w emulsions, a surfactant with an HLB number in the range 8 to 18 (depending on the nature of the oil) is chosen, whereas for w/o emulsions an HLB number in the range 3 to 6 is usually chosen.

The other major application of surfactants is for the preparation of solid/liquid dispersions (usually referred to as suspensions). There are generally two methods for preparing suspensions, referred to as the condensation and dispersion methods. In condensation, the starting point is a molecular unit and the particles are built up by a process of nucleation and growth. A typical example is the preparation of polymer latices. In this case, the monomer (such as styrene or methylmethacrylate) is emulsified in water using an anionic or non-ionic surfactant (such as sodium dodecyl sulphate or alcohol ethoxylate). An initiator, such as potassium persulphate, is added and when the temperature of the system is increased initiation occurs, resulting in the formation of the latex (polystyrene or polymethylmethacrylate). In the dispersion method, preformed particles (usually powders) are dispersed in an aqueous solution containing a surfactant. The surfactant is essential for adequate wetting of the powder (both external and internal surfaces of the powder aggregates and agglomerates must be wetted) [3]. This is followed by dispersion of the powder using high-speed stirrers and finally the dispersion is 'milled' to reduce the particle size to the appropriate range.

In all the above processes, the surfactant has to satisfy the following criteria. For emulsions, the surfactant should be efficient in reducing the interfacial tension at the o/w interface to a low value (this results in reduction of the energy required for emulsification). Usually, the lower the interfacial tension, the smaller the droplets produced. In addition, the surfactant should reduce any coalescence during emulsification by enhancing the Gibbs–Marangoni effect [4, 5]. Due to the incomplete adsorption of the surfactant during emulsification (as a result of the continuous production of a new surface), an interfacial tension gradient results and this produces a dilational elasticity, ε, given by the expression:

$$\varepsilon = \frac{d\gamma}{d\ln A} \tag{8.1}$$

where γ is the interfacial tension and A is the area of the interface. This interfacial tension gradient promotes diffusion of surfactant molecules from the bulk to the interface. These molecules carry solvent molecules with them, resulting in stabilisation of the liquid film between the droplets. This phenomenon of liquid flow as a result of interfacial tension gradients is referred to as the Marangoni effect [4] and it stabilises the emulsion against coalescence during emulsification.

Once the emulsion is produced, the emulsifier has to provide stability against three main processes, namely flocculation, Ostwald ripening and coalescence. Flocculation is reduced or eliminated by providing a strong repulsive force (electrostatic and/or steric repulsion). Ostwald ripening results from the difference in solubility between small and large droplets. The surfactant reduces Ostwald ripening by reducing the interfacial tension and enhancing the Gibbs elasticity [6]. Coalescence during storage is reduced by stabilisation of the liquid film between the droplets. A useful concept is the disjoining pressure, π, introduced by Deryaguin and co-workers [7]. The disjoining pressure π consists of three components:

$$\pi = \pi_{\text{vdw}} + \pi_{\text{el}} + \pi_{\text{s}} \tag{8.2}$$

where π_{vdw} is the disjoining pressure resulting from the van der Waals' attraction (usually negative), π_{el} is the component resulting from double layer repulsion and π_{s} is the component resulting from steric repulsion between the stabilising moieties of the surfactant chains (it may also include a structural component resulting from the orientation of the solvent molecules at the interface).

Additional effects that may contribute to the reduction of coalescence are interfacial viscosity and elasticity [8]. These effects usually arise from the film rigidity produced by polymers and proteins.

For suspensions, the surfactant has also to satisfy a number of criteria, some of them similar to those described above for emulsions. The surfactant has to be a good wetting and dispersing agent. It also has to enhance stabilisation against flocculation and it should reduce Ostwald ripening.

Most of the above criteria for stability are best served by using a polymeric surfactant. In particular, molecules of the A–B, A–B–A block and BA_n graft copolymers (see below) are the most efficient for stabilisation of emulsions and suspensions. In these copolymers the B chain (referred to as the 'anchoring' chain) is chosen to be highly insoluble in the medium and to have a high affinity to the surface (in the case of suspensions) or be soluble in the oil (in the case of emulsions). The A chain is chosen to be highly soluble in the medium and strongly solvated by its molecules. These block and graft copolymers are ideal for preparation of concentrated emulsions and suspensions, which are needed in many industrial applications.

In this chapter, the general classification of polymeric surfactants is described first. This is followed by a discussion of their solution properties. The next section is devoted to the adsorption of polymeric surfactants at the solid/liquid (S/L) interface and a summary of some of the theoretical treatments and methods that may be applied to studying polymeric surfactant adsorption and its conformation at the S/L interface is given. The same principles may be applied to the adsorption of polymeric surfactants at the liquid/liquid (L/L) interface, although theoretical treatments of this problem are not as developed as those for the S/L interface. The next section is devoted to the principles involved in stabilisation of emulsions and suspensions using polymeric surfactants, i.e. the general theory of steric stabilisation. Two sections are devoted to describing the use of polymeric surfactants for the stabilisation of suspensions and emulsions. Finally, a section is devoted to the application of polymeric surfactants for the preparation of stable water-in-oil-in-water (w/o/w) multiple emulsions, which have attracted considerable attention in recent years for application in personal-care products [9] and agrochemical formulations.

8.3 General classification

Perhaps the simplest type of polymeric surfactant is a homopolymer, which is formed from the same repeating units, such as poly(ethylene oxide) or poly(vinyl pyrrolidone). These homopolymers have little surface activity at the o/w interface since the homopolymer segments (ethylene oxide or vinylpyrrolidone) are highly water soluble and have little affinity to the interface. However, such homopolymers may adsorb significantly at the S/L interface. Even if the adsorption energy per monomer segment to the surface is small (a fraction of kT, where k is the Boltzmann constant and T is the absolute temperature), the total adsorption energy per molecule may be sufficient to overcome the unfavourable entropy loss of the molecule at the S/L interface.

Clearly, homopolymers are not the most suitable emulsifiers or dispersants. A variant is to use polymers that contain specific groups that have high affinity to the surface. This is exemplified by partially hydrolysed poly(vinyl acetate) (PVAc), technically referred to as poly(vinyl alcohol) (PVA). The polymer is prepared by partial hydrolysis of PVAc, leaving some residual vinyl acetate groups. Most commercially available PVA molecules contain 4–12% acetate groups. These acetate groups, which are hydrophobic, give the molecule its amphipathic character. On a hydrophobic surface such as polystyrene, the polymer adsorbs with preferential attachment of the acetate groups on the surface, leaving the more hydrophilic vinyl alcohol segments dangling in the aqueous medium. These partially hydrolysed PVA molecules also exhibit surface activity at the o/w interface [10].

The most convenient polymeric surfactants are those of the block and graft copolymer type. A block copolymer is a linear arrangement of blocks of variable monomer composition. The nomenclature for a diblock is poly-A-block-poly-B and for a triblock is poly-A-block-poly-B-poly-A. The most widely used triblock polymeric surfactants are the 'Pluronics' (BASF, Germany) and 'Synperonic PE' (ICI, UK), which consist of two poly-A blocks of poly(ethylene oxide) (PEO) and one block of poly(propylene oxide) (PPO). Several chain lengths of PEO and PPO are available. More recently triblocks of PPO–PEO–PPO ('inverse Pluronics') have become available for some specific applications.

These polymeric triblocks can be applied as emulsifiers or dispersants; the assumption is made that the hydrophobic PPO chain resides at the hydrophobic surface, leaving the two PEO chains dangling in aqueous solution and hence providing steric repulsion.

Although the triblock polymeric surfactants have been widely used in various applications in emulsions and suspensions, some doubt has arisen on how effective they are. It is generally accepted that the PPO chain is not sufficiently hydrophobic to provide a strong 'anchor' to a hydrophobic surface or to an oil droplet. Indeed, the reason for the surface activity of the PEO–PPO–PEO triblock copolymers at the o/w interface may stem from a process of 'rejection' anchoring of the PPO chain since it is not soluble both in oil and water.

Several other di- and triblock copolymers have been synthesised, although these are of limited commercial availability. Typical examples are diblocks of polystyrene-block-polyvinyl alcohol, triblocks of poly(methyl methacrylate)-block-poly(ethylene oxide)-block-poly(methyl methacrylate), diblocks of polystyrene block-polyethylene oxide and triblocks of polyethylene oxide-block-polystyrene-polyethylene oxide [11].

An alternative (and perhaps more efficient) polymeric surfactant is the amphipathic graft copolymer consisting of a polymeric backbone B (polystyrene or polymethyl methacrylate) and several A chains ('teeth'), such as polyethylene oxide [12]. This graft copolymer is sometimes referred to as a 'comb' stabiliser. This copolymer is usually prepared by grafting a macromonomer such methoxy polyethylene oxide methacrylate with polymethyl methacrylate. The 'grafting' technique has also been used to synthesise polystyrene–polyethylene oxide graft copolymers.

8.4 Solution properties

To understand the solution behaviour of polymeric surfactants of the block and graft type it is essential to consider the solution properties of the simpler homopolymers. The behaviour of polymers in solution has been considered in the thermodynamic treatment of Flory [13] and Huggins [14], usually referred to as the Flory–Huggins theory. This theory considers the free energy of mixing pure polymer with pure solvent, ΔG_{mix}, in terms of two contributions, namely the enthalpy of mixing, ΔH_{mix}, and the entropy of mixing, ΔS_{mix}, i.e. the second law of thermodynamics:

$$\Delta G_{mix} = \Delta H_{mix} - T\Delta S_{mix} \tag{8.3}$$

where T is the temperature in K.

Assuming that the polymer chain adopts a configuration on a lattice (provided by solvent molecules) and considering that the mixing is 'random', then the entropy of mixing ΔS_{mix} is given by:

$$\Delta S_{mix} = -k(n_1 \ln \phi_1 + n_2 \ln \phi_2) \tag{8.4}$$

where k is the Boltzmann constant, n_1 is the number of solvent molecules with a volume fraction ϕ_1 and n_2 is the number of polymer molecules with a volume fraction ϕ_2.

The enthalpy of mixing, ΔH_{mix}, is given by:

$$\Delta H_{mix} = n_1 \phi_2 \chi kT \tag{8.5}$$

where χ is a dimensionless interaction parameter and χkT expresses the difference in energy of a solvent molecule in pure solvent compared to its immersion in pure polymer. χ is usually referred to as the Flory–Huggins interaction parameter.

Combining equations (8.3) to (8.5):

$$\Delta G_{mix} = kT(n_1 \ln \phi_1 + n_2 \ln \phi_2 + \chi n_1 \phi_2) \tag{8.6}$$

The mixing of a pure solvent with a polymer solution creates an osmotic pressure, π, which can be expressed in terms of the polymer concentration c_2 and the volume fraction of the polymer:

$$\frac{\pi}{C_2} = RT\left[\frac{1}{M_2} + \left(\frac{v_2^2}{V_1}\right)\left(\frac{1}{2} - \chi\right)c_2 + \ldots\right] \tag{8.7}$$

where v_2 is the partial specific volume of the polymer ($v_2 = V_2/M_2$), V_1 is the molar volume of the solvent and M_2 is. the molecular weight of the polymer.

The second term in equation (8.7) is the second virial coefficient B_2, therefore:

$$\frac{\pi}{c_2} = RT\left(\frac{1}{M_2} + B_2 + \ldots\right) \tag{8.8}$$

$$B_2 = \left(\frac{v_2^2}{V_1}\right)\left(\frac{1}{2} - \chi\right) \tag{8.9}$$

Note that $B_2 = 0$ when $\chi = ½$, i.e. the polymer behaves as ideal in mixing with the solvent. This condition was termed by Flory as the θ point [13]. Under these conditions the polymer chains in solution have no repulsion or attraction and they adopt their unperturbed dimension. Clearly when $\chi < ½$, B_2 is positive and mixing is non-ideal, leading to positive deviation (repulsion); this occurs when the polymer chains are in 'good' solvent conditions. In contrast, when $\chi > ½$, B_2 is negative and mixing is non-ideal, leading to negative deviation (attraction); this occurs when the polymer chains are in 'poor' solvent conditions (precipitation of the polymer may occur under these conditions).

Since the polymer solvency depends on temperature, a theta temperature, θ, at which $\chi = ½$ can also be defined.

The function $[(½) - \chi]$ can also be expressed in terms of two mixing parameters, an enthalpy parameter κ_1 and an entropy parameter ψ_1, i.e.:

$$\left(\frac{1}{2} - \chi\right) = \kappa_1 - \psi_1 \tag{8.10}$$

The θ temperature can also be defined in terms of κ_1 and ψ_1:

$$\theta = \frac{\kappa_1 T}{\psi_1} \tag{8.11}$$

Alternatively:

$$\left(\frac{1}{2} - \chi\right) = \psi_1\left(1 - \frac{\theta}{T}\right) \tag{8.12}$$

Although the Flory–Huggins theory is sound in principle, several experimental results cannot be accounted for. For example, it was found that the χ parameter depends on the polymer concentration in solution. Most serious is the fact that many polymer solutions (such as PEO) show phase separation on heating, when the theory predicts that it should happen only on cooling.

The solution properties of copolymers are much more complicated. This is due to the fact that the two copolymer components A and B behave differently in different solvents. Only when the two components are both soluble in the same solvent will they exhibit similar solution properties. This is the case, for example, for a non-polar copolymer in a non-polar solvent. It should also be emphasised that the Flory–Huggins theory was developed for ideal linear polymers. Indeed, with branched polymers consisting of high monomer density (such as star branched polymers), the θ temperature depends on the length of the arms and is in general lower than that of a linear polymer with the same molecular weight.

Another complication arises from specific interaction with the solvent, e.g. hydrogen bonding between polymer and solvent molecules (e.g. with PEO and PVA in water). Aggregation in solution (lack of complete dissolution) may also present a problem.

One of the most useful parameters for characterising the conformation of a polymer in solution is the root mean square (rms) end-to-end length $<r^2>^{1/2}$, which represents the configuration character r as the distance from one end group to the other of a chain molecule. Another useful parameter is the radius of gyration $<s^2>^{1/2}$, which is a measure of the effective size of a polymer molecule (it is the root mean square distance of the elements of the chain from its centre of gravity).

For linear polymers:

$$<s^2>^{\frac{1}{2}} = \frac{<r^2>^{\frac{1}{2}}}{6^{\frac{1}{2}}} \tag{8.13}$$

The radius of gyration of a polymer in solution can be determined from light-scattering measurements.

As mentioned above, dilute solutions of copolymers are solvents that are good for both components exhibit similar behaviour to hompolymer chains. However, in a selective solvent, i.e. where the medium is a good solvent for one component, say A, and a poor solvent for the second component B, one part of the amphipathic block or graft will separate as a distinct phase, while the other stays in solution. The insoluble portion of the amphipathic copolymer will aggregate reversibly to form micelles. It is believed that the polymeric micelles are spherical [15]. The critical micelle concentration of these block and graft copolymers is usually very low.

Several methods may be applied to obtain the micellar size and shape of block and graft copolymers, of which light scattering, small-angle X-ray and neutron scattering are probably the most direct. Dynamic light scattering (photon correlation spectroscopy) can also be applied to obtain the hydrodynamic radius of the micelle. This technique is relatively easy to perform when compared with static light scattering since dynamic light scattering does not require rigorous preparation of the samples.

8.5 Adsorption and conformation of polymeric surfactants at interfaces

Understanding the adsorption and conformation of polymeric surfactants at interfaces is vital to knowing how these molecules act as stabilisers. Most basic ideas about the adsorption and conformation of polymers have been developed for the solid/liquid interface [16–18]. The same concepts may be applied to the liquid/liquid interface, with some modification whereby some parts of the molecule may reside within the oil phase rather than simply staying at the interface. Such modification does not alter the basic concepts, particularly when dealing with stabilisation by these molecules.

The process of polymer adsorption involves a number of various interactions that must be separately considered. Three main interactions must be taken into account, namely the interaction of the solvent molecules with the surface (or oil in the case of o/w emulsions), which needs to be displaced for the polymer segments to adsorb, the interaction between the chains and the solvent and the interaction between the polymer and the surface. Apart from these interactions, one of the most fundamental considerations is the conformation of the polymer molecule at the interface.

These molecules adopt various conformations, depending on their structure. The simplest case to consider is that of a homopolymer that consists of identical segments (e.g. poly(ethylene oxide)), which shows a sequence of loops, trains and tails, as illustrated in Figure 8.1a. It should be mentioned at this stage that for such a polymer to adsorb the reduction in entropy of the chain as it approaches the interface must be compensated by the energy of adsorption between the segments and the surface. In other words, the chain segments must have a minimum adsorption energy, χ^s, otherwise no adsorption occurs. With polymers that are highly water soluble, such as poly(ethylene oxide), the interaction energy with the surface may be too small for adsorption to occur, and if this is the case the whole molecule may not be strongly adsorbed on the surface. For this reason, many commercially available polymers that are described as hompolymers, such as poly(vinyl alcohol), contain some hydrophobic groups or short blocks (vinyl acetate in the case of PVA) that ensure their adsorption to hydrophobic surfaces. This is illustrated in Figure 8.1b. Clearly, if all the segments have a high affinity to the surface, the whole molecule may lie flat on the surface, as illustrated in Figure 8.1c. This situation is rarely the case since the molecule will have very low solubility in the continuous medium.

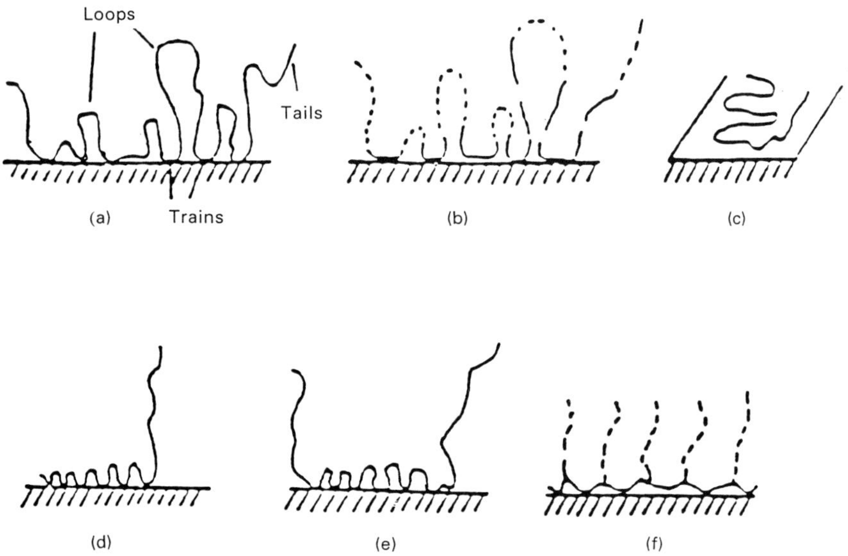

Figure 8.1 Various conformations of polymeric surfactants adsorbed on a plane surface. (a) Random conformation of loops, trains and tails; (b) preferential adsorption of short blocks; (c) chain lying flat on the surface; (d) AB block copolymer with loop-train configuration of B and long tail of A; (e) ABA block as in (d); (f) BA_n graft with backbone B forming small loops and several tails of A ('teeth').

The most favourable structures for polymeric surfactants are those shown in Figures 8.1d, e and f, referred to as block and graft copolymers. The molecule shown in Figure 8.1d is an A–B block, consisting of a B chain that has a high affinity for the surface (or is soluble in the oil phase), referred to as the 'anchoring' chain, and an A chain that has very low affinity for the surface and is strongly solvated by the medium. As will be discussed in the section on stabilisation, this is the most convenient structure since the forces that ensure strong adsorption are opposite to those that ensure stability. A variant on the structure shown in Figure 8.1d is the A–B–A block copolymer shown in Figure 8.1e. In this case, the anchor chain B contains two stabilising chains (tails). Another variant is shown in Figure 8.1f, which is described as a graft copolymer ('comb'-type structure) with one B chain and several A chains (tails or 'teeth').

It is clear from the above description of polymer configurations that for full characterisation of the process of adsorption it is necessary to know the following parameters: the amount of polymer adsorbed per unit area of the surface, Γ (mol/m^2 or mg/m^2), the fraction of segments in close contact with the surface, p, and the distribution of polymer segments, $\rho(z)$, from the surface towards the bulk solution. It is also essential to know how far the segments extend into solution, i.e. the adsorbed layer thickness δ. It is important to establish how these parameters change with polymer coverage (concentration), the structure of the polymer and its molecular weight. It is also essential to know how these parameters change with the environment, i.e. solvency of the medium for the chains and temperature.

Several theories exist that describe the process of polymer adsorption. These have been developed using either the statistical mechanical approach or quasi-lattice models. In the statistical mechanical approach the polymer is considered to consist of three types of structures with different energy states: trains, loops and tails [16]. The structures close to the surface (trains) are adsorbed with an internal partition function determined by short-range forces between the segment and surface (assigned an adsorption energy per segment χ^s). The segments in the loops and tails are considered to have an internal partition function equivalent to that of segments in bulk solution and these are assigned a segment–solvent interaction parameter χ (Flory-Huggins interaction parameter). By equating the chemical potential of the macromolecule in the adsorbed state and in bulk solution, the adsorption isotherm can be determined. In the earlier theories, the case of an isolated chain on the surface (low coverage) was considered, but later the theories were modified to take into account the lateral interaction between the chains, i.e. at high coverage.

The quasi-lattice model was developed by Roe [17] and Scheutjens and Fleer [18]. The basic procedure is to describe all chain conformations as step-weighted random walks on a quasi-crystalline lattice that extends in parallel layers away from the surface. This is illustrated in Figure 8.2, which shows a possible conformation of a polymer molecule at a surface.

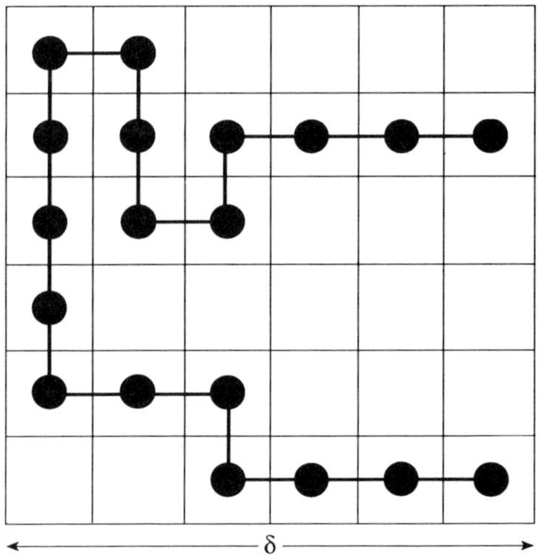

Figure 8.2 Possible conformation of a polymer molecule at an interface.

The partition function is written in terms of the number of chain configurations, which are treated as connected sequences of segments. In each layer random mixing (Bragg–William or mean field approximation) between segments and solvent molecules is assumed. Each step in the random walk is assigned a weighting factor p_i that is considered to consist of three contributions: the adsorption energy χ^s, the configurational entropy of mixing and the segment–solvent interaction parameter χ.

The above theories give a number of predictions for polymeric surfactant adsorption, which are summarised as follows. As the number of segments in the chain increases from a low (with few segments) to a high (large number of segments) value, the adsorption isotherm changes from a Langmuirian type (characteristic for surfactant adsorption) to a high affinity type. In the latter case, the first addition of polymer chains to the solution results in their virtually complete adsorption. Adsorption in this case is described as being 'irreversible', i.e. the equilibrium between adsorbed and free polymer is shifted towards the surface. This explains the strong anchoring of the polymer chains to the surface. As the solvency of the medium for the chains decreases, the amount of polymer adsorbed increases. This explains the relatively 'weaker' adsorption of homopolymers that are highly solvated by the medium. It is now clear from these theories why block and graft copolymers are preferred for the stabilisation of dispersions. The poor solubility of the anchor chain B in the medium and its strong affinity to the surface ensures the strong adsorption of the molecule. In contrast, the high solubility of the stabilising chain A ensures effective steric stabilisation.

Another prediction from the theories is that the higher the molecular weight of the polymer, the higher the amount of adsorption, when the latter is expressed in mg/m^2. The bound fraction, p, increases with increasing adsorption energy, χ^s, but it decreases with increasing surface coverage and increasing molecular weight of the polymer. The segment density distribution, $\rho(z)$, and the adsorbed layer thickness δ increase with increasing molecular weight of the polymer. It also increases with increasing solvency of the medium for the chains.

Several experimental methods can be applied to study the adsorption and conformation of polymers at interfaces. The amount of polymer adsorbed, Γ, can be directly determined by equilibrating a known amount of the disperse phase (particles or droplets) with a polymer solution of known concentration, C_1. When the system reaches equilibrium (this may take hours or even days with high molecular weight polymers), the disperse phase is separated (by filtration or centrifugation) and the equilibrium concentration of the polymer, C_2, is determined using a suitable analytical method. From C_1 and C_2, the amount of disperse phase m, and its surface area A (in m^2/g, which may be obtained from a knowledge of the particle size distribution), Γ can be calculated:

$$\Gamma = \frac{C_1 - C_2}{mA} \tag{8.14}$$

As an illustration of this, Figure 8.3 shows the adsorption isotherms at 25°C for poly(vinyl alcohol) (containing 12% acetate groups) on polystyrene latex [19]. The figure shows the high affinity isotherms for the polymers and the increase in adsorption of the polymer with increasing molecular weight. Similar isotherms are expected for the adsorption of the polymer on oil droplets. However, in this last case it is not possible to obtain the full isotherm since to produce the emulsion requires a minimum amount of polymer. In addition, the surface area of the emulsion has to be determined at each point from the droplet size distribution.

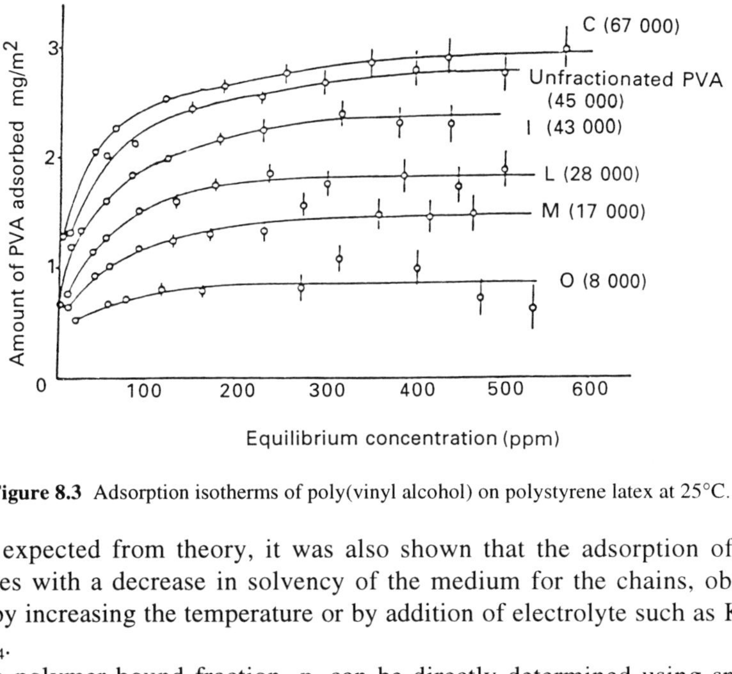

Figure 8.3 Adsorption isotherms of poly(vinyl alcohol) on polystyrene latex at 25°C.

As expected from theory, it was also shown that the adsorption of PVA increases with a decrease in solvency of the medium for the chains, obtained either by increasing the temperature or by addition of electrolyte such as KCl or Na_2SO_4.

The polymer-bound fraction, p, can be directly determined using spectroscopic methods such as NMR. The method depends on the reduction in the mobility of the segments that are in close contact with the surface. By using a pulsed NMR technique p can be estimated. An indirect method for the estimation of p is the use of microcalorimetry. Basically the enthalpy of adsorption per molecule is compared with that per segment [20]. The latter may be obtained by using small molecules of similar structure to a polymer segment.

Several methods may be used to determine the adsorbed layer thickness, δ. Most of the methods depend on measuring the hydrodynamic radius of the particles with and without the adsorbed polymer layer. For example, the relative viscosity, η_r, of a dispersion with an adsorbed polymer layer may be measured. Assuming that the particles behave as hard spheres (when δ is small compared with the particle radius R) of non-interacting units (low volume fraction of the disperse phase), η_r can be related to the effective volume fraction, ϕ_{eff}, by the Einstein equation:

$$\eta_r = 1 + 2.5\phi_{eff} \tag{8.15}$$

From ϕ_{eff} and ϕ (the core volume fraction), the adsorbed layer thickness, δ_h, can be calculated using the equation:

$$\phi_{eff} = \phi\left(1 + \frac{\delta_h}{R}\right)^3 \tag{8.16}$$

In order to apply the above method a dispersion of monodisperse particles with a radius that is not much larger than δ_h should be used. This limits the method for direct use with emulsions, which are polydisperse and have relatively large radii. However, small model particles of polystyrene may be used and the assumption is made that the polymer conformation at the interface is similar to that at the o/w interface.

Another method for determining δ_h is to apply dynamic (quasi-elastic) light scattering, referred to as photon correlation spectroscopy (PCS). By measuring the intensity fluctuation of scattered light (using a laser) by the particles, as a result of their Brownian diffusion, the diffusion coefficient, D, of the particles can be determined [20]. It is necessary to use small particles (which show Brownian motion) with a radius that is not much larger than δ_h. From D the hydrodynamic radius R_h of the particles can be obtained using the Stokes–Einstein equation:

$$D = \frac{kT}{6\pi\eta R_h} \tag{8.17}$$

By carrying out the measurements in the presence and absence of polymer layers δ_h can be obtained. In the presence of an adsorbed polymer layer the hydrodynamic radius is the sum of the core radius and the adsorbed layer thickness, whereas in the absence of the polymer layer, R_h is simply the core radius. Again, as with the viscosity technique, this method cannot be directly applied to emulsions, which are polydisperse and have a radius that is large for significant Brownian motion.

8.6 Interaction between particles (droplets) containing adsorbed polymer layers (steric stabilisation)

When two particles or droplets containing adsorbed polymer layers (with an adsorbed layer thickness δ) approach a distance of separation h whereby these layers begin to overlap, i.e. when $h < 2\delta$, repulsion occurs as a result of two main effects. The first repulsive force arises from the unfavourable mixing of the polymer layers when these are present in a good solvent (i.e. the chains are strongly solvated by the medium). The unfavourable mixing of polymer solutions in good solvent conditions was considered by Flory and Krigbaum [20], whose theory was applied to the case of interparticle interaction. A schematic representation of the mixing of polymer layers on close approach is shown in Figure 8.4, which shows the situation when two particles with polymer layers are forced to approach to a distance h that is less than 2δ, forming an overlap region with a volume element dV. Before overlap, the chains have a volume fraction ϕ_2 and the solvent has a chemical potential μ_1^{α}. In the overlap region, the volume fraction of the chains is ϕ_2', which is higher than ϕ_2, and the solvent has a chemical potential μ_1^{β}, which is lower than μ_1^{α}. This is equivalent to an increase in the osmotic pressure in the overlap region. As a result, solvent diffuses from the bulk to the overlap region and the two particles or droplets are separated, i.e. strong repulsion results. The latter is referred to as mixing or osmotic repulsion.

Using the Flory–Krigbaum theory [20], the free energy of mixing, G_{mix}, resulting from this unfavourable overlap can be calculated:

$$\frac{G_{\text{mix}}}{kT} = \frac{4\pi}{3V_1}\phi_2^2 N_{\text{av}}\left(\frac{1}{2} - \chi\right)\left(\delta - \frac{h}{2}\right)^2\left(3R + 2\delta + \frac{h}{2}\right) \tag{8.18}$$

where k is the Boltzmann constant, T is the absolute temperature, V_1 is the molar volume of the solvent and N_{av} is Avogadros' constant.

It is clear from equation (8.18) that when the Flory–Huggins interaction parameter, χ, is less than 0.5, i.e. the chains are in good solvent conditions, G_{mix} is positive, the interaction is repulsive and it increases very rapidly with decreasing h when the latter is lower than 2δ. This explains why polymeric surfactants such as Hypermer CG6 (a graft copolymer of polymethyl methacrylate backbone and PEO side chains, produced by ICI) is ideal for stabilising dispersions in aqueous media. For stabilisation of dispersions in non-aqueous media, such as w/o emulsions, the stabilising chains have to be soluble in the oil phase (normally a hydrocarbon). In this case polyhydroxystearic acid (PHS) chains are effective. A polymeric surfactant such as Aralacel P135 (an ABA block copolymer of PHS–PEO–PHS produced by ICI) is an ideal w/o emulsifier.

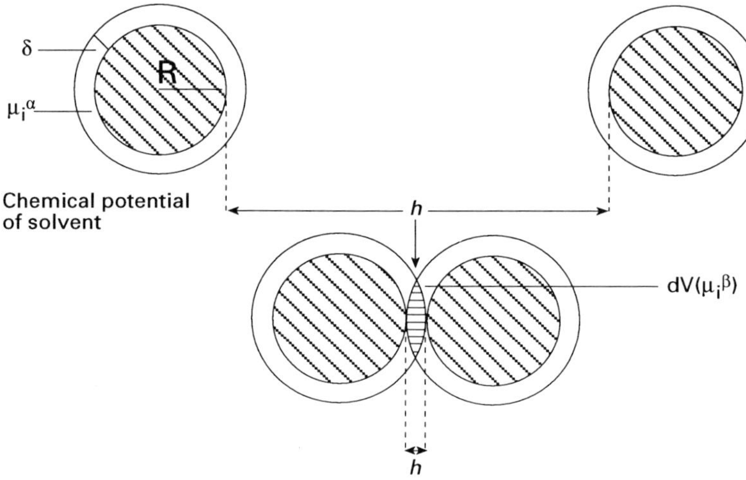

Figure 8.4 Schematic representation of the overlap of two polymer layers.

Equation (8.18) also shows that when $\chi > 0.5$, i.e. when the solvency of the medium for the chains becomes poor, G_{mix} is negative and the interaction becomes attractive. The condition $\chi = 0.5$ is referred to as θ solvent and this denotes the onset of the change from repulsion to attraction. To ensure steric stabilisation by the above mechanism it must be ensured that the chains are kept in better than a θ solvent.

The second repulsive force arises from the loss of configurational entropy when the chains overlap. This is schematically illustrated in Figure 8.5, where the polymer chain is represented by a simple rod with one attachment point on the surface. When the two surfaces are separated at infinite distance, each chain will have a number of configurations, Ω_∞, that are determined by the volume of the hemisphere swept by the rod. When the two surfaces approach to a distance h that is smaller then the radius of the hemisphere swept by the rod, the volume available to the chains becomes smaller and this results in a reduction in the configurational entropy to a value Ω (which is less than Ω_∞). This results in strong repulsion and the effect is referred to as entropic, volume restriction or elastic repulsion and is given by:

$$G_{el} = 2\nu \ln \frac{\Omega}{\Omega_\infty} \tag{8.19}$$

where ν is the number of polymer chains per unit area of the surface. It should be mentioned that G_{el} is always repulsive and it becomes very high on considerable overlap of the polymer chains.

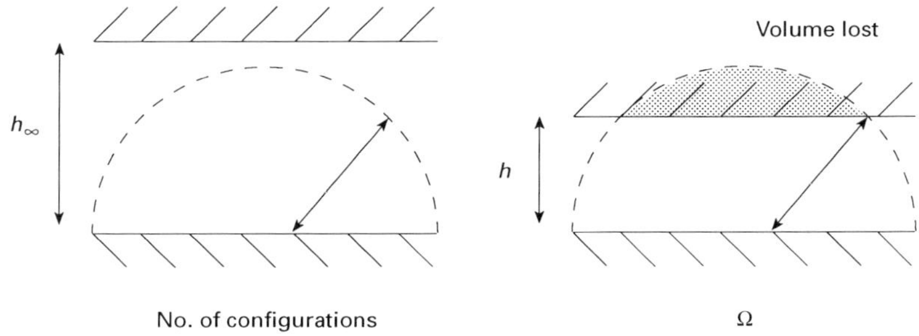

Figure 8.5 Schematic representation of the entropic, volume restriction or elastic interaction.

Plots of G_{mix} and G_{el} versus h are illustrated in Figure 8.6. This figure shows that G_{mix} increases very rapidly with a decrease in h as soon as h becomes smaller than 2δ (and $\chi < 0.5$). G_{el} also increases very rapidly with a decrease in h on further overlap. Combination of G_{mix}, G_{el} and G_A (the van der Waals' attraction) results in the total $G_T - h$ curve shown in Figure 8. 6. This curve shows a minimum (G_{min}) at $h \cong 2\delta$, but when $h < 2\delta$, G_T increases very rapidly with further decrease in h. The depth of G_{min} depends on the adsorbed layer thickness. With increasing δ, G_{min} decreases and at sufficiently high values of δ (of the order of 5–10 nm), it reaches small values (fraction of kT units). This shows that, with sterically stabilised dispersions, there is only weak attraction at relatively long distances of separation, which in most cases is overcome by Brownian diffusion. Thus, the net interaction is repulsive and this ensures the long-term stability of the dispersion.

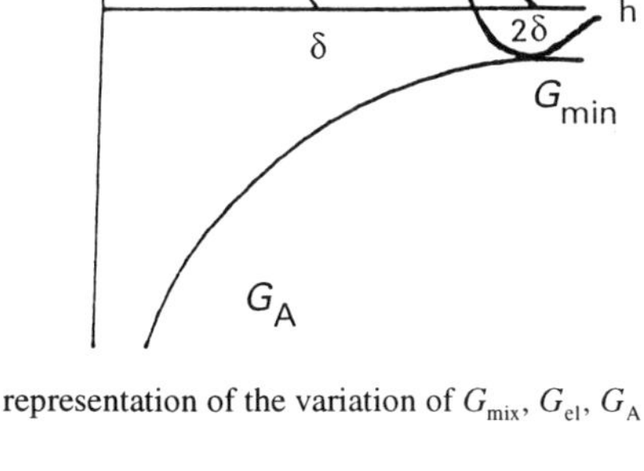

Figure 8.6 Schematic representation of the variation of G_{mix}, G_{el}, G_A and G_T with h.

From the above discussion the main criteria for effective steric stabilisation can be summarised. First, there should be enough polymer to ensure complete coverage of the surface by the chains. This will prevent any attraction between the bare patches or bridging by the polymer chains (which can adsorb simultaneously on more than one particle). Second, the chains must be strongly adsorbed (anchored) to the surface. This prevents any displacement on close approach. In this respect block and graft copolymers containing an anchoring chain (such as polystyrene or poly(methyl methacrylate) to hydrophobic surfaces) are the best stabilisers. In some cases strong adsorption may be achieved by a phenomenon referred to as rejection anchoring. If the B chain is insoluble in the medium (water or oil) it can adsorb by this rejection mechanism. This situation applies to poly(propylene oxide), which is insoluble in water and in most oils. This explains the use of block copolymers of PEO–PPO–PEO as emulsifiers.

The third criterion for effective steric stabilisation is to ensure that the stabilising chain A remains in good solvent conditions at all times and under all conditions. As discussed above, for systems where water is the continuous medium PEO is the most suitable A chain. The polymer chain is highly soluble in water and remains solvated up to high temperatures. It can also tolerate reasonable amounts of electrolyte. For dispersions, where the continuous medium is a hydrocarbon oil (e.g. w/o emulsions), polyhydroxystearic acid is the most suitable A chain.

The last criterion for effective steric stabilisation is to have a sufficiently thick adsorbed layer to avoid any weak flocculation. This is particularly important for concentrated dispersions. As discussed above, a value of δ of 5–10 nm is usually sufficient. Hence the side PEO chains need to have a molecular weight of the order of 1000–2000. With most graft copolymers these side chains are extended and they provide a sufficiently thick layer.

8.7 Polymeric surfactants for suspension stabilisation

Several block and graft copolymers have been developed for stabilisation of suspensions of hydrophobic particles in aqueous media. The adsorption of two graft copolymers, namely Atlox 4913 and Hypermer CG-6(a) (supplied by ICI Surfactants), on polystyrene latex has been studied recently [21]. Both copolymers have the same backbone, namely poly(methyl methacrylate–methacrylic acid). The Hypermer CG-6(a) has a higher proportion of polymethacrylic acid. The side chains are the same, namely methoxy-polyethylene glycol (MePEG). By increasing the polymethacrylic acid content in the backbone, the PEG chain density decreases, i.e. the Hypermer CG-6(a) will have a lower side chain density than Atlox 4913. Typical adsorption isotherms for the two graft copolymers on two latices (with diameters of 867 and 427 nm) at 20°C are shown in Figure 8.7. The adsorption isotherms are of the Langmuir type and the plateau adsorption values are the same on both latices, indicating that the particle curvature has no effect on the amount of adsorption. However, the plateau adsorption value is higher for Atlox 4913 compared to the value for Hypermer CG-6(a) (≈1.5 and ≈1.2 mg/m^2, respectively). This may be due to the fact that the copolymer side chain density is lower for Hypermer CG-6(a) when compared to that of Atlox 4913.

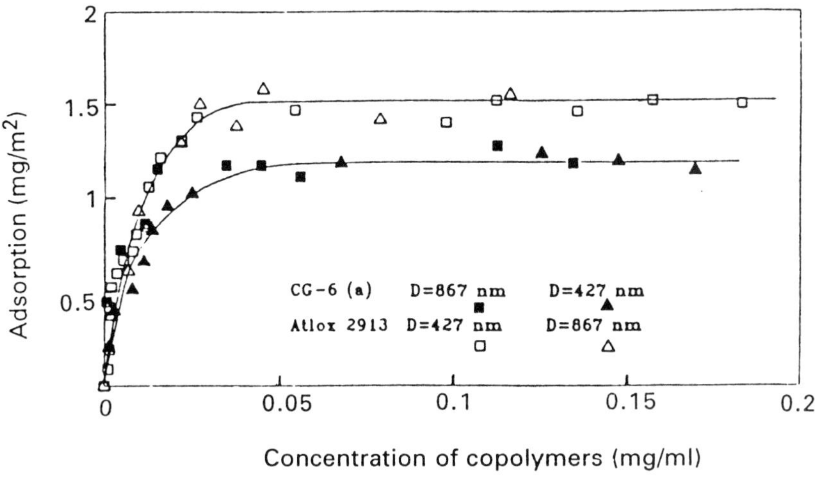

Figure 8.7 Adsorption isotherms for graft copolymers on polystyrene latex at 20°C.

The adsorbed layer thickness of the two graft copolymers was determined using rheological measurements, as described before [21–24]. Basically the relative viscosity is determined as a function of the volume fraction of the suspension and the data fitted to a hard sphere model, as suggested by Dougherty and Krieger [25], i.e.:

$$\eta_r = \phi\left[1-\left(\frac{\phi_{eff}}{\phi_m}\right)\right]^{-[\eta]\phi_m} \qquad (8.20)$$

where ϕ_m is the maximum packing fraction, which can be obtained from a plot of $(\eta_r^{1/2}-1)/\phi$ versus $(\eta_r^{1/2})$ in which the slope is equal to $1/\phi_m$. ϕ_{eff} is the effective volume fraction, which can be used to obtain δ_h using equation (8.16). The adsorbed layer thickness for Atlox 4913 on the 427 nm latex was 18 nm at $\phi=0.37$ and decreased to 9.6 nm when the volume fraction was increased to 0.50. This is attributed to the interpenetration and/or compression of the chains with increasing volume fraction, as discussed before [24–26].

Similar trends were obtained for the adsorption of the above graft copolymers on TiO_2 but in this case the adsorption of Hypermer CG-6(a) was higher because of the more hydrophilic nature of the TiO_2 surface.

One of the best methods for studying the interaction of the above sterically stabilised dispersions is the application of dynamic rheological measurements. In this method a small amplitude sinusoidal strain is applied on the dispersion (placed in the gap of a concentric cylinder platen) while simultaneously measuring the stress. Form the amplitudes of the stress and strain (τ_o and γ_o respectively) and the phase angle shift between the sine waves of stress and strain, δ, the complex modulus, G^*, the storage modulus, G' (the elastic component of the complex modulus) and the loss modulus, G'' (the viscous component of the complex modulus) can be obtained:

$$|G^*| = \frac{\tau_o}{\gamma_o} \tag{8.21}$$

$$G' = |G^*| \cos \delta \tag{8.22}$$

$$G'' = |G^*| \sin \delta \tag{8.23}$$

As an example of this, Figure 8.8 shows log–log plots of G^*, G' and G'' (at a frequency of 0.1 Hz) as a function of the volume fraction of the latex suspension, ϕ_s, with a diameter of 867 nm. The cross-over point at which $G' = G''$ occurs at $\phi_s = 0.42$. This cross-over point, ϕ_{cr}, can be taken as the point where the adsorption layers just begin to overlap. Below ϕ_{cr}, $G'' > G'$ and the interaction between the particles is weak. However, above ϕ_{cr} $G' > G''$ and the adsorbed layers may interpenetrate or compress. Under these conditions, the system becomes predominantly elastic [21]. Similar trends were obtained with the smaller latex but in this case ϕ_{cr} was lower (0.31). This is what is expected since in this case the ratio of adsorbed layer thickness to particle radius is larger.

The rheological results (see Figure 8.8) illustrate the strong steric repulsion obtained using the graft copolymers as soon as the PEO chains begin to overlap. With small particles, such overlap occurs at relatively low solid volume fractions. As the particle size increases, the volume fraction at which overlap begins increases and with coarse suspensions (of a radius of 1 μm or higher) it will be close to the maximum packing fraction (0.6 or higher). This allows the preparartion of highly concentrated suspensions that are very stable in the colloid sense. The strong adsorption of the graft copolymers and the high solvation of the PEO chains ensures that the stability is maintained over a long period of time. This explains why such graft copolymers are very effective in the stabilisation of concentrated suspensions. In addition, the viscosity of such highly concentrated suspensions remains low up to high volume fractions. For example, with agrochemical suspension concentrates (where the average particle radius is between 1 and 2 μm), the high shear viscosity of the system could be maintained below 100 mPa.s at volume fractions of 0.6. These suspensions could be formulated to final products in which a gelling agent and a surfactant could be incorporated.

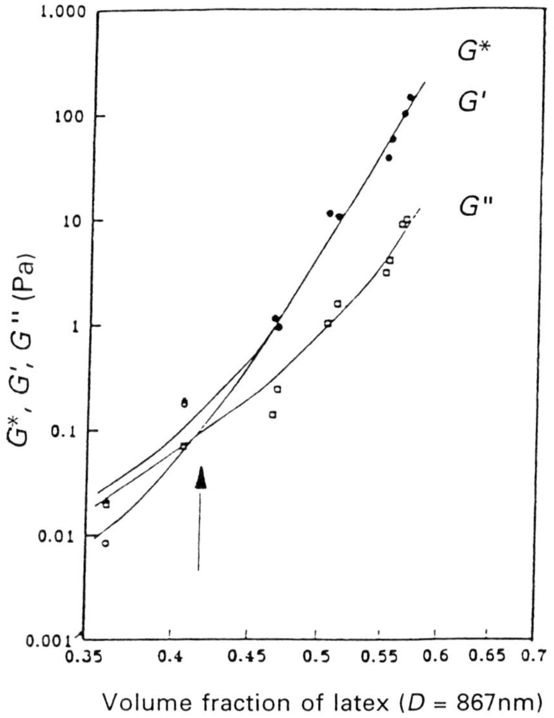

Figure 8.8 Plot of moduli (at 0.1 Hz) as a function of latex (D=867 nm, stabilised with Atlox 4913) volume fraction.

8.8 Polymeric surfactants as stabilisers for emulsions

Several block copolymers can be used for the stabilisation of oil-in-water and water-in-oil emulsions [26]. Synperonic NPE 1800 and its analogues (supplied by ICI Surfactants) are ideal for stabilisation of o/w emulsions in which the oil is non-polar (e.g. paraffinic oils). These surfactants have the following structural formula:

$$C_9H_{19}\text{–}C_6H_4\text{–}(CH_2\text{–}CH(CH_3)\text{–}O)_m\text{–}(CH_2\text{–}CH_2\text{–}O)_n\text{–}OH$$

They all contain the same hydrophobic group (nonyl phenol and 13 moles of propylene oxide) but have different numbers of moles of ethylene oxide (27 for Synperonic NPE 1800, 48 for Synperonic NPE A, 80 for Synperonic NPE B and 174 for Synperonic NPE C). For example, Figure 8.9 shows the viscosity–volume fraction curves for paraffin oil-in-water emulsions stabilised using the Synperonic NPE surfactants. The average droplet diameter for each emulsion was determined using the Coulter counter and the volume mean diameter (VMD) is given in the caption. For comparison, relative viscosity was calculated using the Dougherty–Krieger equation (equation (8.20)). It can be seen from Figure 8.9 that the emulsions behave very nearly as hard sphere dispersions. This is not surprising since the adsorbed layer thickness (5.8–11.6 nm) is much smaller than the particle radius. The emulsions prepared using these polymeric surfactants were very stable against coalescence.

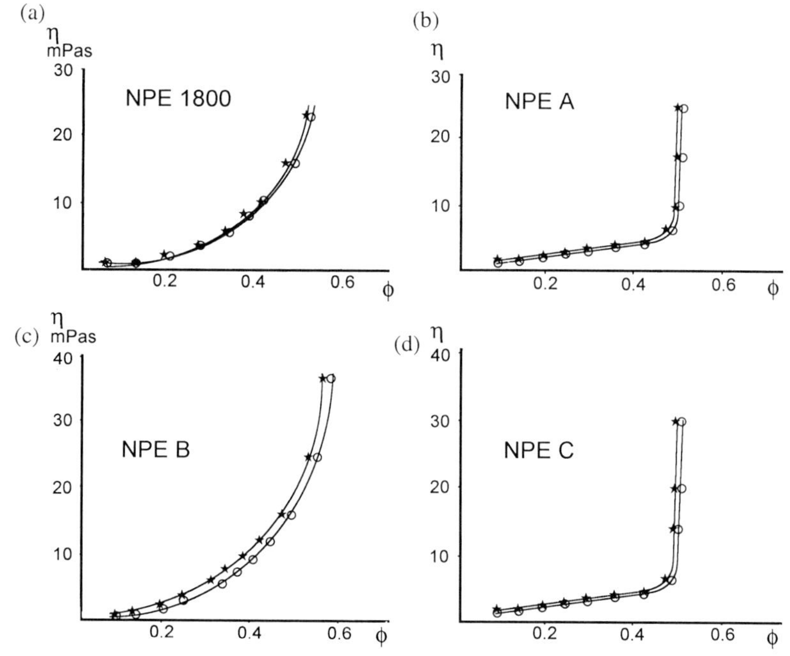

Figure 8.9 η–ϕ curves for paraffin oil/water emulsions (x = experimental results; o = theoretical values according to equation (8.6). (a) Synperonic NPE 1800, VMD = 3.5 μm; (b) NPE A, VMD = 4 μm; (c) NPE B, VMD = 4.5 μm; (d) NPE C, VMD = 5 μm.

Block copolymers of the ABA type could also be used to prepare very stable w/o emulsions. In this case the B chain (the anchoring chain) has to have high affinity to the water droplets, whereas the A chains have to be soluble in the oil and strongly solvated by the oil molecules. An example of such a block copolymer is

Arlacel P135 (supplied by ICI Surfactants). This consists of a B chain of poly-(ethylene oxide) and two A chains of polyhydroxystearic acid (PHS–PEO–PHS). The relative viscosity–volume fraction curve for water/isoparaffinic oil emulsions is shown in Figure 8.10. The emulsion has a narrow size distribution with an average z radius of 183 nm. Using equation (8.20), the effective volume fraction was calculated and the results are included in Figure 8.10. From ϕ_{eff}, the adsorbed layer thickness was calculated and this was found to be ≈ 10 nm at $\phi = 0.4$, decreasing to ≈ 4 nm at $\phi = 0.65$. This reduction in δ with increasing ϕ may be attributed to the interpenetration and/or compression of the chains on increasing ϕ.

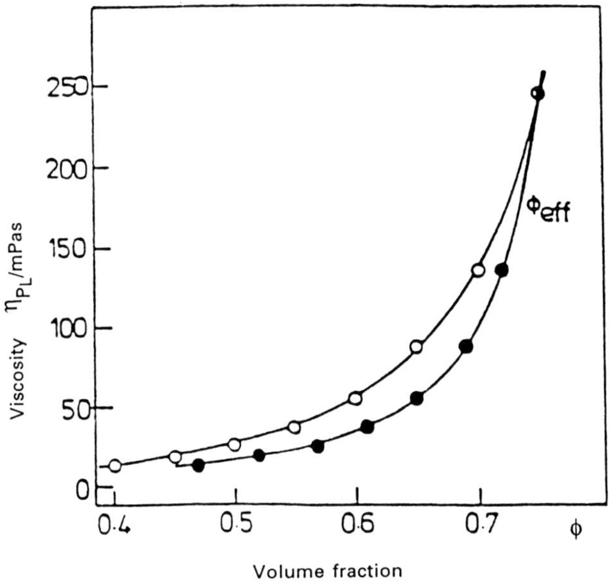

Figure 8.10 Viscosity–volume fraction curves for w/o emulsions stabilised with Arlacel P135.

Using the above ABA block copolymer, highly concentrated w/o emulsions can be prepared and these emulsions are stable for a long period of time (more than a year) at high temperatures (up to 50°C). This illustrates the advantage of using polymeric surfactants for stabilisation of w/o emulsions and these can find use in cosmetic creams (see below).

8.9 Polymeric surfactants in multiple emulsions

As discussed before [9], multiple emulsions (w/o/w or o/w/o) are ideal systems for applications in cosmetics for the following reasons. First, ingredients can be dissolved in three different compartments. For example, with w/o/w multiple emulsions, two different water-soluble additives (proteins, enzymes and vitamins) and an insoluble ingredient (e.g. perfume) can be incorporated. Second, these systems can be used for sustained release by control of the breakdown process on application. Third, they allow the same cream consistency to be produced for many cosmetics, e.g. by incorporation of a thickener or gelling agent in the outer continuous phase.

The main criteria for the preparation of stable multiple emulsions are summarised as follows [9]. First, two optimum emulsifiers are required, one with a low and one with a high HLB number. Emulsifier 1 (with a low HLB number), which is used to prepare the primary w/o emulsion (for the case of w/o/w multiple emulsion), should prevent any flocculation or coalescence of the emulsion droplets. This emulsifier should ideally produce a viscoelastic film at the w/o interface that ensures the stability of the emulsions and prevents any transport of components from the internal droplets to the outer continuous medium during storage. Recent work has shown that Aralacel P135 satisfies these criteria [27]. It produces a viscoelastic film and the w/o emulsions prepared using this polymeric surfactant are stable over a long period of time and at various temperatures. The secondary emulsifier, with a high HLB number, should also produce an effective barrier to prevent flocculation and coalescence of the resulting multiple emulsion drops. Again a polymeric surfactant, such as Synperonic PF 127 (an ABA block of PEO–PPO–PEO produced by ICI) is ideal in this case.

The second criterion for stabilisation of the multiple emulsion on storage is to have an optimum osmotic balance between the internal (water droplets) and the external continuous medium. This can be achieved by the use of electrolytes or non-electrolytes. Usually, the osmotic pressure in the external continuous medium is kept slightly lower than that of the internal water droplets.

The multiple emulsion is prepared by a two-stage process, as illustrated in Figure 8.11. The primary emulsion is prepared by adding the aqueous phase to an oil solution of the polymeric emulsifier 1 with a low HLB number, e.g. Aralcel P135, using a high-speed stirrer. Using this polymeric emulsifier, primary water droplets that are smaller than 1 μm in diameter can be produced. The size of the droplets in the primary emulsion can be determined using photon correlation spectroscopy. This primary emulsion is then emulsified into an aqueous solution of the high HLB polymeric surfactant (Synperonic PEF 127) using a low-speed stirrer (paddle stirrer) to produce multiple emulsion drops around 10–100 μm in diameter.

A schematic representation of the multiple emulsion drop is shown in Figure 8.12, which illustrates the various components. In some cases a polymer coating (produced using, for example, polymethacrylic acid) may be used to ensure the long-term physical stability of the multiple emulsion. Using polymeric surfactants multiple emulsions that are stable for more than a year can be produced [28]. By incorporating a gelling agent such as Carbopol in the aqueous continuous phase, creaming of the multiple emulsion could be prevented. This gelling agent also produces the right consistency (rheology) for application as a cream.

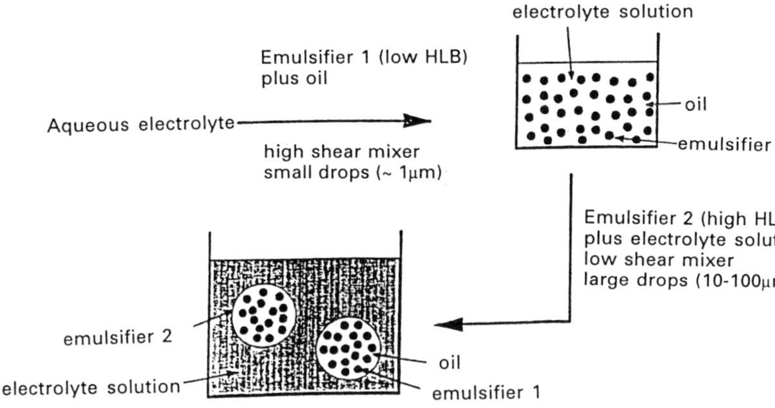

Figure 8.11 Schematic representation of the preparation of w/o/w multiple emulsion.

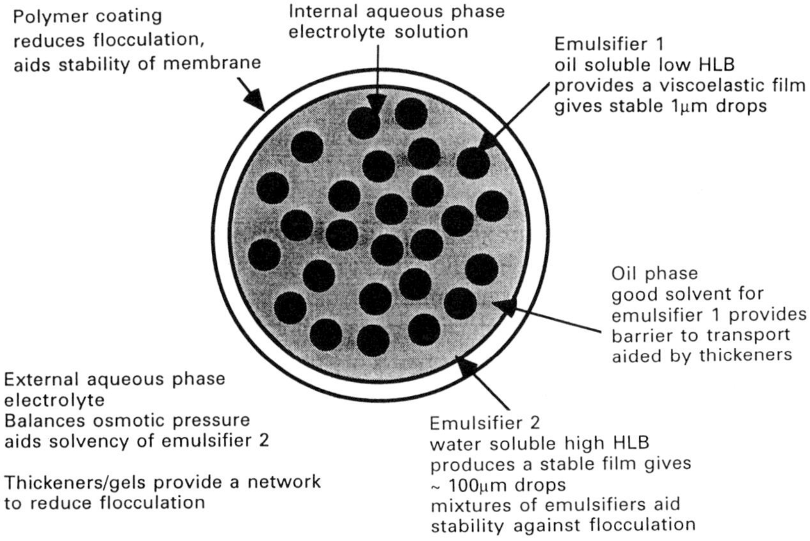

Figure 8.12 Schematic representation of the multiple emulsion drop showing the role of the various components.

References

[1] Griffin, W.C. (1949) *J. Cosmetics Chemists*, **1** 311.
[2] Griffin, W.C. (1954) *J. Cosmetics Chemists*, **5** 249.
[3] Walstra, P. (1983) in *Encyclopedia of Emulsion Technology*, vol.1 (ed. P. Becher), Marcel Dekker, New York.
[4] Tadros, Th.F. (1987) *Solid/liquid Dispersions*, Academic Press, London.
[5] Tadros, Th.F. and Vincent, B. (1983) in *Encyclopedia of Emulsion Technology*, vol.1 (ed. P. Becher), Marcel Dekker, New York.
[6] Vincent, B. (1984) in *Surfactants* (ed. Th.F. Tadros), Academic Press, London.
[7] Deryaguin, B.V. and Obucher, E. (1935) *J. Colloid Chem.*, **1** 385.
[8] Tadros, Th.F. (1994) *Colloids and Surfaces*, **91** 39.
[9] Tadros, Th.F. (1992) *Int. J. Cosmet. Sci.*, **14** 93.
[10] Lankfeld, J. and Lyklema, J. (1972) *J. Colloid Interface Sci.*, **44** 454, 465, 475.
[11] Piirma, I. (1992) *Polymeric Surfactants*, Marcel Dekker, New York.
[12] Paxton, T.R. (1969) *J. Colloid Interface Sci.*, **31** 19.
[13] Flory, P.J. (1941) *J. Chem. Phys.*, **9** 660; **10** 51.
[14] Huggins, M.L. (1942) *J. Chem. Phys.*, **46** 151.
[15] Reiss, G., Hurtrez, G. and Badahur, P. (1985) *Encyclopedia of Polymer Science and Engineering*, Wiley, New York, p. 324.
[16] Silberberg, A. (1968) *J. Chem. Phys.*, **48** 2835.
[17] Roe, R.J. (1974) *J. Chem. Phys.*, **60** 4192.
[18] Scheutjens, J.M.H.M. and Fleer, G.J. (1982) *Adv. Colloid Interface Sci.*, **16** 341.
[19] Garvey, M.J., Tadros Th.F. and Vincent, B. (1974) *J. Colloid Interface Sci.*, **49** 57.
[20] Flory, P.J. (1953) *Principles of Polymer Chemistry*, Cornell University Press, New York,
[21] Liang, W., Bognolo G. and Tadros, Th.F. (1995) *Langmuir*, **11** 2899.
[22] Liang, W., Tadros, Th. F. and Luckham, P.F. (1992) *J. Colloid Interface Sci.*, **153** 131.
[23] Prestidge, C. and Tadros, Th.F. (1988) *J. Colloid Interface Sci.*, **124** 660.
[24] Pingret, F.J.V., Sohm, R.H. and Tadros, Th.F. (1992) *Colloids and Surfaces*, **65** 85.
[25] Krieger, I.M. (1972) *Adv. Colloid Interface Sci.*, **3** 111.
[26] Th.F. Tadros and P.D. Winn, to be published.
[27] A. Lucel and Th.F. Tadros, to be published.
[28] Py, C., Rouviere, J., Loll, P., Taelman, M.C. and Tadros, Th.F. (1994) *Colloids and Surfaces*, **91** 215.

9 The formulation and industrial applications of multiple emulsions: an area of fast development

A. Silva Cunha, J.L. Grossiord and M. Seiller

9.1 Introduction

Multiple emulsions are systems in which the drops of dispersed phase themselves contain smaller droplets that are miscible with the external continuous phase. Thus the multiple emulsion may be of the water-in-oil-in-water (w/o/w) type, where the internal and external aqueous phase are separated by an oil phase, or of the oil-in-water-in-oil (o/w/o) type,where the aqueous phase is surrounded by two oil phases.

Multiple emulsions can be viewed as systems that control the transport of molecules from an external to an internal phase or vice versa. These emulsions could be of interest in many industrial processes, for example the extraction of phenol in waste waters or the separation of hydrocarbons from seawater after an oil spillage, etc. However, the major potential applications are in pharmaceutical and cosmetic formulations.

As often happens the therapeutic interest of multiple emulsions was discovered by accident. Before the Second World War, a researcher was administering a vaccine to some test animals. To facilitate the administration he thought of dispersing the emulsion in the water, just before the injection. Not only was the administration easier (as expected), but an additional surprising effect was noted: all the animals treated with the dispersed preparation showed a rate of antibody higher than that observed with animals immunised with the simple water-in-oil (w/o) emulsion formula. The explanation of the phenomenon is that after dispersion in water of the simple w/o emulsion a light agitation is sufficient to form a w/o/w multiple emulsion in the syringe. Thus, the release of the antigen incorporated in the internal aqueous phase is slowed down, which causes an increase in its efficiency. This first accidental utilisation of multiple emulsions in therapy was followed by many others in different areas.

Thanks to their structure multiple emulsions have interesting properties that are useful in many applications. The liquid oily or aqueous intermediate membrane of these emulsions allows the incorporation of a high level of different substances in the internal phase (compared to other systems) enabling:

- two incompatible substances to be made available at the same time: one in the inner phase and one in the outer phase

- complete protection of the substance encapsulated
- differentiated drug-releases, i.e. one drug is made immediately available by inclusion in the external phase while there is a delayed release for the drug incorporated in the internal phase
- possible drug targeting because of the vesicular character of the system
- extraction and separation processes.

9.2 Formulation

The formulation, characterisation and industrial production of multiple emulsions are well established [1].

9.2.1 Nature of the oil phase

In the field of pharmaceutical, cosmetic and food preparations, the oil phases used most frequently to form w/o/w or o/w/o multiple emulsions are hydrocarbons [2], triglycerides and esters (Table 9.1).

Among the hydrocarbons, liquid paraffin is quite popular because of its low toxicity and ease of emulsification.

Triglycerides are used for *in vivo* tests because they are well tolerated by the organism. Their emulsification in aqueous phases, however, is difficult.

Isopropyl myristate is the most widely used ester. It is also possible to obtain stable multiple emulsions with fatty acids [4], fatty alcohols [17] and various silicones [18]. However, these substances are generally used to adjust the viscosity of the oil phase.

Nowadays more and more semi-synthetic oils are used, such as various medium-chain triglycerides (MCT). These oils are lighter than conventional oils and give emulsions with an oily external phase that is less viscous and greasy.

The nature of the oil has a great influence on the characteristics of the system, and physico-chemical properties such as polarity, density and viscosity determine the behaviour of multiple emulsions [19].

9.2.2 Nature of the emulsifier

Non-ionic emulsifiers are preferred (Table 9.2) in pharmaceutical, cosmetic and food formulations because of their good emulsion stabilisation properties and their relatively low toxicity. Among the non-ionic emulsifiers the most common are the esters of fatty acids with a chain length of 18 carbons (saturated or unsaturated). Non-ionic emulsifiers such as fatty alcohol ethoxylates are less popular because of their less favourable toxicological profile. The natural emulsifiers, e.g. phospholipids and casein–dextran complexes, are rarely used because of their low emulsification efficiency, which does not compensate entirely for their excellent toxicology and skin tolerance.

Table 9.1 Oils and additives used in the formulation of multiple emulsions

Oils	Additives	References
Soybean oil	Polyvinylpyrrolidone	[3]
	Gelatin	
	Arabic gum	
Oleic acid	—	[4]
Penasete® 800[a]	Gelatin	[5]
	Amino acids	
Liquid paraffin	Hydroxy propyl methyl cellulose	[6]
Cetiol S®[b]	Magnesium sulfate	[7]
Arachis oil	Gelatin	[8]
Olive oil	Arabic gum	
	Cherry gum	
Isopropyl myristate	BSA	[9]
	Gelatin	
	Polyacrylic acid	
	Polyvinyl alcohol	
Soybean oil	Gelatin	[10]
Liquid paraffin	Polyvinyl alcohol	[11]
	Polyvinylpyrrolidone	
	Soya lecithin	
	Arabic gum	
	Calcium stearate	
Squalene	—	[12]
Isopropyl myristate	—	[13]
Lipiodol®[c]		
Triglycerides 5545®[d]	Sodium sulfate	[14]
Soybean oil	Sodium chloride	[15, 16]
Medium-chain triglycerides		

[a]Penasete® 800 = glyceryl tricaprylate.
[b]Cetiol® = dioctylcyclohexane.
[c]Lipiodol® = iodised poppy-seed oil ethyl ester.
[d]Triglycerides 5545® = caprilic/capric triglycerides.

Multiple emulsions are sometimes produced with polymeric emulsifiers such as Poloxamers® (polyoxyethylene–polyoxypropylene block copolymers), Abil® (polysiloxane polyalkyl polyether copolymer), Hypermer® (ABA block copolymers of polyhydroxystearic acid and polyethylene oxide) [27], fatty acid–tetraglyceryl condensates, etc. Nevertheless, despite the interest in these new polymeric emulsifiers, classic emulsifiers are most commonly used because their properties are well known and their toxicology is proven.

Table 9.2 Emulsifiers used in multiple emulsions

Lipophilic emulsifier[a]	Hydrophilic emulsifier[b]	References
Span® 80 (1)	Tween® 40 (1)	[20]
Span® 80 (2)	Tween® 80 (2)	[21]
Lecithin	HCO®-60 (3)	
	Brij® 35 (4)	
	Pluronic® F68 and F88 (5)	
Arlacel® C (3)	HCO®-60	[22]
	Pluronic® F68	
Span® 80 (2)	Tween® 80	[6]
Dehymuls® A (4)	Emulgin® SMO 20 and SMS 20 (6)	[7]
	Mergital® E1212 (7)	
Hypermer® A60 (5)	Synperonic® PE/F127 (8)	[23]
Tetraglyceryl-condensated ricinoleate	Poloxamer 188 (9)	[24]
Hexaglyn® PR-15 (6)	HCO®-60	[25]
Hexaglyceryl-condensed ricinoleate	Casein–dextran complex	[26]
Span® 80 (2)	Block copolymers	[12]
HCO®-40 (7)	Pluronic® F68	[13]
Abil® EM90 (8)	Brij® 58 (10)	[14]
	Generol® 122 E-25 (11)	
Abil® EM90 (8)	Tween® 80 (2)	[15, 16]

[a]Lipophilic emulsifiers: (1) sorbitan mono-oleate; (2) sorbitan sesquioleate; (3) mixture of sorbitan esters; (4) modified polyester, (5) polymeric polyester; (6) hexaglyceryl condensed polyricinoleate; (7) derivatives of hydrogenated castor oil; (8) polysiloxane polyalkyl/polyether copolymer.

[b]Hydrophilic emulsifiers: (1) polyoxyethylene sorbitan monopalmitate; (2) polyoxyethylene sorbitan mono-oleate; (3) derivatives of hydrogenated castor oil; (4) polyoxyethylene lauryl ether; (5) poloxamer compounds (polyoxyethylene–polyoxypropylene copolymers); (6) polyoxyethylene sorbitan mono-oleate and monostearate; (7) POE olyelalcohol; (8) block copolymer of ethylene oxide; (9) POE-20 cetyl alcohol; (10) PEG-25 soya sterol.

To formulate a w/o/w multiple emulsion at least two emulsifiers are introduced to the system: one lipophilic, to form the primary w/o emulsion, and the other hydrophilic, to form the final multiple emulsion. The hydrophilic–lipophilic balance (HLB) of the primary emulsifier is usually in the range 3–7 while the secondary emulsifier has an HLB of between 8 and 16. The HLB value for the total emulsifier system must generally be below 11 [28]. If the value is higher, there is a risk of phase inversion and the formation of an o/w emulsion. The HLB value of the emulsifier mixture can be calculated using Frenkel's equation [29]:

$$\text{weighted HLB} = \frac{(\text{HLB}_1 \times \Phi_{\text{wow}} \times W_1) + (\text{HLB}_2 \times W_2)}{\Phi_{\text{wow}} \times (W_1 + W_2)}$$

where HLB_1 and HLB_2 are the HLB values of the primary and the secondary emulsifiers, respectively, W_1 and W_2 are the weight ratios of the primary and secondary emulsifiers to the total surfactant concentration, respectively, and Φ_{wow} is the volume fraction of the w/o emulsion in the multiple emulsion.

9.2.3 Optional additives

Hydrophilic macromolecules, such as bovine serum albumin (BSA), amino acids, carboxyvinylic compounds and cellulose and its analogues, introduced in the internal aqueous phase may increase the stability of the system (Table 9.1). The increased stability can be attributed to a complexation between macromolecule and emulsifier at the water–oil interface.

Similarly, thickening agents, like gelatine, xanthan gum, guar gum, arabic gum, carboxyvinylic compounds, cellulose and its analogues and polyvinylpyrrolidone, can be added to increase the viscosity and consequently the stability of the emulsions. It is also possible to enhance stability by incorporation of aluminium salts and fatty acids in the oil phase.

9.2.4 Nature of the encapsulated substances

Substances that can be included in the internal phase are active species or markers. The latter include mainly electrolytes [16, 30, 31], sugars, dyes [32] and fluorescent agents [33]. Electrolytes (NaCl, $MgSO_4$) are most often used:

- to control the osmotic gradient between the internal and external aqueous phases
- to increase the stability of systems because of salting-out effects
- to follow up the stability/instability of the emulsion and the rate of breakdown.

9.3 Methods of preparation

Several processes, simple and complex, are reported in the literature.

9.3.1 The two-step process

This is the most widely used procedure [16, 34]. The first step consists of preparing a primary emulsion. The second step involves dispersing a given amount of the primary emulsion in the external phase containing the secondary emulsifier (Figure 9.1). Although the method is simple, the second step of the process includes a critical emulsification phase.

Similarly to simple emulsions, each parameter in the emulsification process has an influence on the characteristics of the final system, but this influence is more pronounced in the case of multiple emulsions. In the manufacture of a w/o/w multiple emulsion the first emulsification step requires typical w/o emulsion conditions, i.e. high manufacturing temperature (about 70°C) and a high speed of agitation (generally above 1500 rpm with a turbine agitator) for about 25 to 35 min and then until completely cooled.

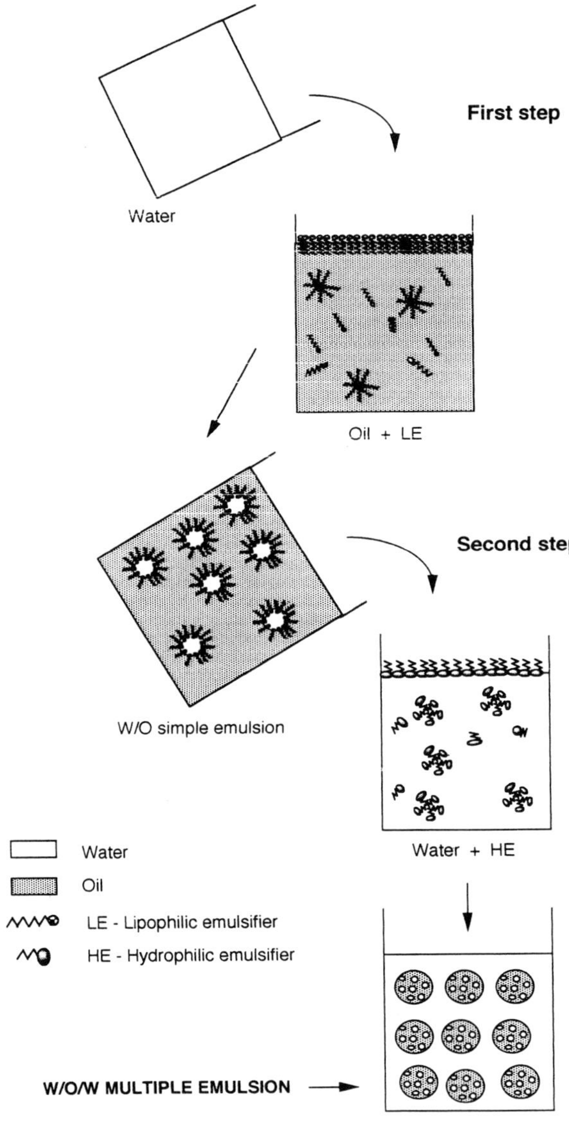

Figure 9.1 Preparation of water-in-oil-in-water multiple emulsions by the two-step process.

During the second emulsification step, in order to avoid the breakdown of the oil droplets, some precautions have to be taken. The introduction of the primary emulsion in the aqueous emulsifier solution is better done at room temperature and should be slow and progressive (about 60 min), and the speed of agitation should be relatively low and generally less than 1500 rpm with a turbine agitator. It is also recommended to continue stirring for 20 min after complete addition of the primary emulsion to thc cxtcrnal aqucous phasc. However, it is interesting to note that it is possible to prepare a multiple emulsion with a high speed of agitation and a short mixing time (only a few minutes) but this method is more difficult to control.

The main advantage of the two-step process is that it is a fully controlled procedure which should, at least in theory, allow the quantity of internal phase to be controlled. Its disadvantage is its limited reproducibility because the second emulsification is a critical step. In the case of a w/o/w multiple system, the primary w/o emulsion usually has a high viscosity and is difficult to disperse. During this step, there is the risk of breakdown of the oil droplets, causing the mixing of part of the internal phase with the external water phase.

The two-step process is the method most often described and employed for the preparation of o/w/o multiple emulsions [35, 36].

9.3.2 The phase-inversion process

This procedure is also performed in two steps and is quite similar to the two-step emulsification method [37], the only difference being the addition of the external phase to the primary emulsion (as opposed to the incorporation of the primary emulsion into the external phase) (Figure 9.2). The advantage of this process is that it is very easy to perform and gives a better control of the internal phase volume than the two-step process. It has the disadvantage of requiring two emulsification steps and, more importantly, of giving sometimes simple instead of multiple emulsions. Indeed, in the case of a w/o/w multiple emulsion a very slight excess of water is sufficient to transform the multiple emulsion into a simple emulsion of the o/w type. Conversely, an incorporation of water that is too slow causes the formation of a simple emulsion of the w/o type. The two-step and phase-inversion processes, despite some problems, can be conveniently used to produce multiple emulsions on an industrial scale. The two methods described below represent novel developments in the preparation of w/o/w multiple emulsions.

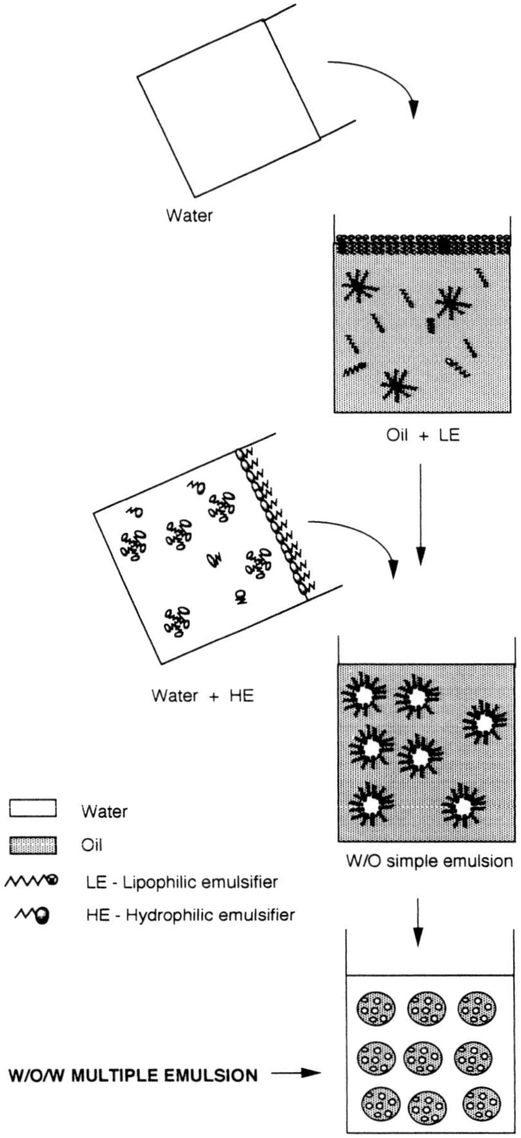

Figure 9.2 Preparation of water-in-oil-in-water multiple emulsions by the phase-inversion process.

9.3.3 The lamella-phase dispersion process

This method has been optimised and used with success by Vigie [38]. It originates from the process used to prepare liposome-like vesicles with non-ionic emulsifiers. It can be used only when the components form a lamellar phase when mixed with water in certain proportions (Figure 9.3).

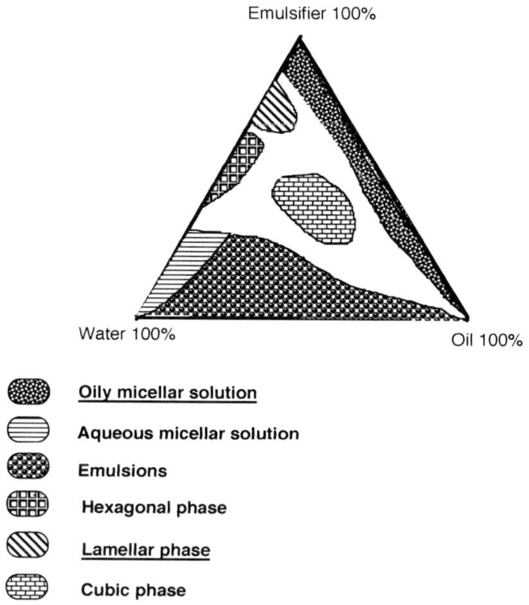

Figure 9.3 Phase diagram obtained (HLB ≃ 11) with liquid paraffin and ethoxylated oleyl alcohol mixture.

The lamellar phase is prepared by mixing at 70°C about 60% emulsifier, 10% oil and 30% demineralised water containing an electrolyte. The dispersion is made using, for example, a turbine agitator at 500 rpm for 10 min. To the mesophase thus obtained a given amount of water is added progressively, at a very high speed of agitation (>1500 rpm), until formation of a w/o/w multiple emulsion is achieved (Figure 9.4).

This procedure has the advantage of requiring only a single emulsification step. The mesophase formed by an optimal ratio of lipophilic and hydrophilic emulsifiers in water is thermodynamically stable and can be obtained rapidly and easily. The main limitation of this process is that not all emulsifiers give a lamellar phase. When this mesophase exists, the HLB of the total emulsifier system is often high, which is disadvantageous for the stability of a multiple emulsion. In addition, the quantity of oil that can be incorporated in the lamellar phase is rarely higher than 10%. Another disadvantage of this process is the limited control of the yield of encapsulation of active substances.

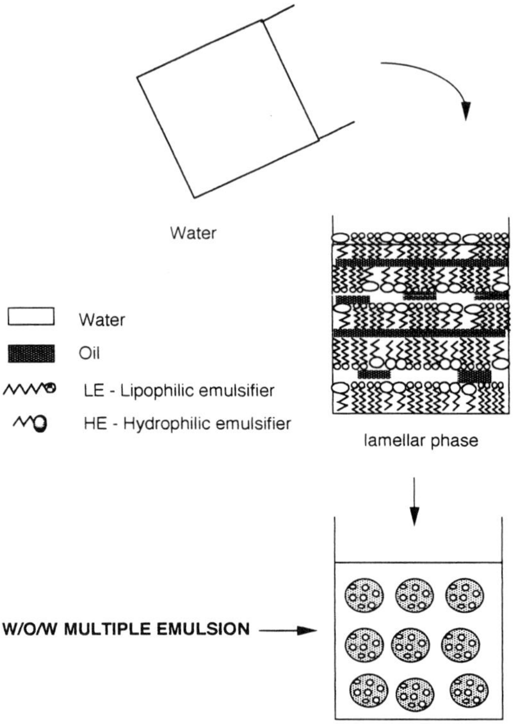

Figure 9.4 Preparation of water-in-oil-in-water multiple emulsions by the lamellar-phase dispersion process.

9.3.4 The oil-isotrope dispersion process

This method has been tested by Vigie [38]. Initially, the oily micellar isotrope (Figure 9.3) is prepared at 70°C to 80°C by mixing for 30 min at about 1000 rpm 50% of emulsifiers' mixture, 40% oil and 10% demineralised water containing an electrolyte. Water is then gradually introduced to this phase, usually with a turbine agitator at room temperature for about 30 min but at a lower speed (200 to 400 rpm) (Figure 9.5).

The preparation of a multiple emulsion by dispersion of an oily isotrope in water offers in principle the same advantages as the lamellar-phase dispersion process, i.e. only one emulsification step is required. The initial phase is a pure phase, stable and easy to obtain. The main disadvantage is the low level of water solubilised in the inverse micelles, which rarely exceeds 10%.

These two processes, lamellar-phase dispersion and oil-isotrope dispersion, require large quantities of emulsifier. Water is added to the lamellar phase or to the oil isotrope, which is not really in the form of droplets, and therefore it is difficult to control the total amount of water in the internal phase.

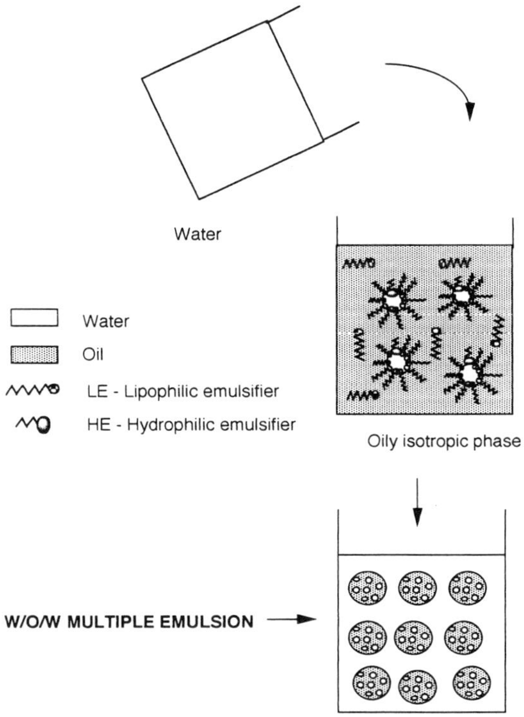

Figure 9.5 Preparation of water-in-oil-in-water multiple emulsion by the oily-isotrope dispersion process.

9.3.5 Other methods

Kavaliunas *et al.* describe a process involving the mixing of various phases (an oil-rich solution of emulsifier, L2, an aqueous micellar solution of emulsifier, L1, and a mesomorphous phase, N) [39]. However, this process is rarely used because it is necessary to have, in the same emulsion, various emulsifiers capable of producing the three phases.

Pilman *et al.* have prepared a system with an inverse micellar solution consisting of water and monolaurate or tetraglycerol monolaurate and soybean oil [40]. The system obtained is emulsified with an aqueous solution of sodium caseinate. According to these authors this method allows the preparation of w/o/w multiple-microemulsions.

Makino *et al.* describe a one-step mechanical process with continuous agitation [41]. In this process equal quantities of an aqueous phase containing a small quantity of a hydrophilic emulsifier and of an oil phase containing a high proportion of a lipophilic emulsifier are mixed. The equipment used is specific and provides high agitation at constant speed. The preparation of multiple emulsions by this method is not common and depends on a large quantity of lipophilic emulsifier, the rate of flow of the external liquid phase, the viscosity of the two liquids, the mixing ratio of the two phases and, above all, the specific design of the emulsifying equipment.

Higashi *et al.* describe a new method for the preparation of w/o/w multiple emulsions by the membrane-emulsification technique [23]. This method allows the formation of monodispersed lipid microdroplets containing aqueous microdroplets to form a w/o/w system. This procedure has been used, with some modifications, in industrial-scale processes.

9.4 Characterisation

The main assays for the characterisation of multiple emulsions are:

- *Macroscopic analysis.* For this assay, the organoleptic properties of the multiple emulsions are evaluated.
- *Microscopic analysis.* This assay gives information about the size of the particles, the multiple character of the emulsions and the ratio between simple and multiple droplets (Figure 9.6).
- *Granulometric analysis.* This analysis is carried out with a particle counter. A small amount of emulsion is diluted in an isotonic solution and about 50 000 particles are analysed to determine the average size of the multiple droplets.

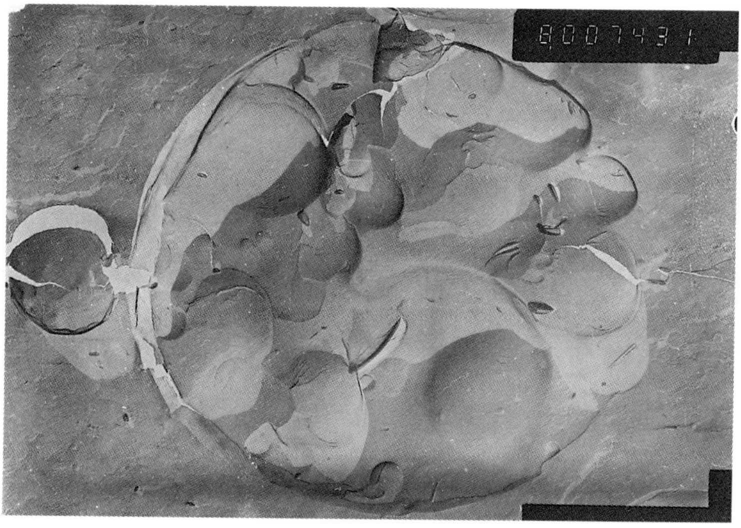

Figure 9.6 Microscopic aspect of a w/o/w multiple emulsion.

- *Thermal stability test.* This test is carried out at 4±1°C, 25±1°C and 40±1°C to observe the homogeneity of the emulsion and the development of any phase separation.
- *Conductimetric test.* This assay is carried out in order to define the yield of the multiple emulsions immediately after preparation. The yield is defined as the ratio (w/w) of the marker present in the internal and the external aqueous phase. This assay is also used to follow emulsion stability on storage.
- *Rheological analysis.* Rheological measurements give a lot of important information about the characteristics of multiple emulsions. Steady-state flow analysis is used to study the behaviour of multiple emulsions under a steady shear rate and to follow the change of viscosity over time (Figure 9.7). In oscillatory experiments a sinusoidal stress applied to the sample gives the basic viscoelastic parameters such as the shear modulus and the phase angle of the stress with respect to the strain. By recording the evolution of these parameters with increasing stress, a fingerprint of the multiple emulsion is obtained.
- *Differential scanning calorimetry (DSC).* The application of this analysis to w/o/w multiple emulsions allows the calculation of the mass fraction of the dispersed and continuous aqueous phases. From the DSC data (solidification enthalpies and temperatures of the aqueous phases) it is therefore possible to find evidence for possible mass transfers between the internal and external phases in a multiple emulsion (Figure 9.8).

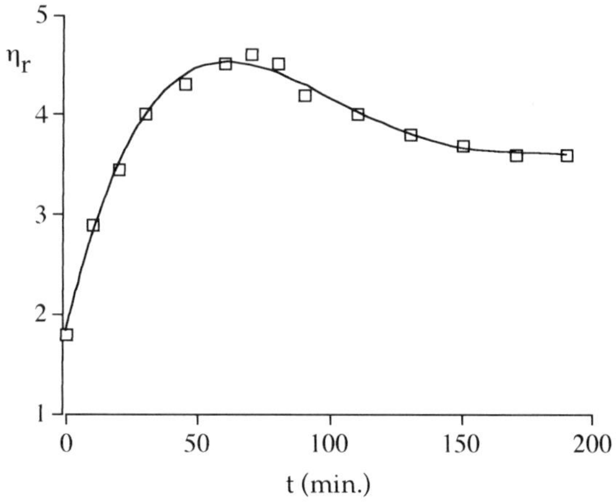

Figure 9.7 Rheological behavior of a w/o/w multiple emulsion during the swelling–breakdown transition.

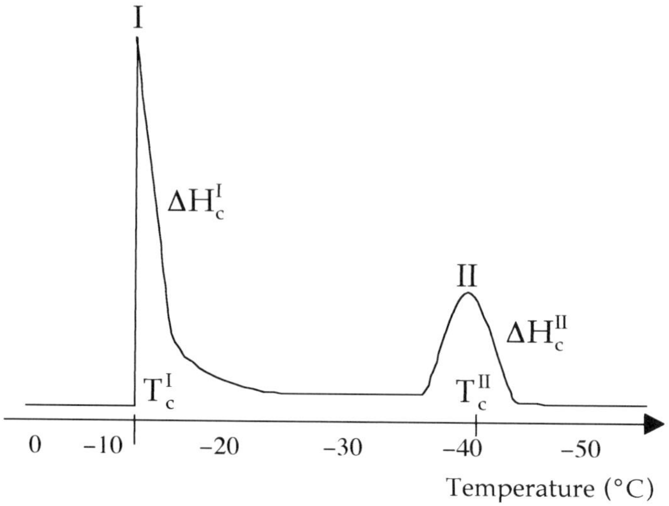

Figure 9.8 DSC thermogram showing the freezing of a w/o/w multiple emulsion sample.

9.5 Mechanisms of transport of the encapsulated substances

The understanding of the mechanisms of transport of water-soluble molecules through the oil membrane between the external and internal aqueous phases is of fundamental importance for the practical application of w/o/w multiple emulsions. These mechanisms can be exploited in many processes, e.g. release, extraction, etc.

There are essentially three mechanisms of transport:

- *passive or simple diffusion:* this process involves hydrosoluble molecules that have sufficient affinity for the oil to allow them to cross this phase.
- *facilitated diffusion by a carrier:* in the case where the molecules do not have sufficient affinity for the oil, a passage through the oily membrane with the assistance of a carrier (diffusion within micelles or diffusion after formation of a lipo-soluble complex) can be envisaged.
- *release after breakdown of the oil droplets:* the breakdown can take place by application of a strong shear or by dilution of the multiple emulsion under hypo-osmotic conditions. In the second case, the concentration gradient between all the dissolved species is responsible for water flow from the external phase to the internal phase. This aqueous transport produces an increase in the internal droplet size and therefore the oil droplets swell until a critical size is reached. Beyond this critical size, the droplets burst following the breakdown of the oil membrane and release the encapsulated substances (Figure 9.9).

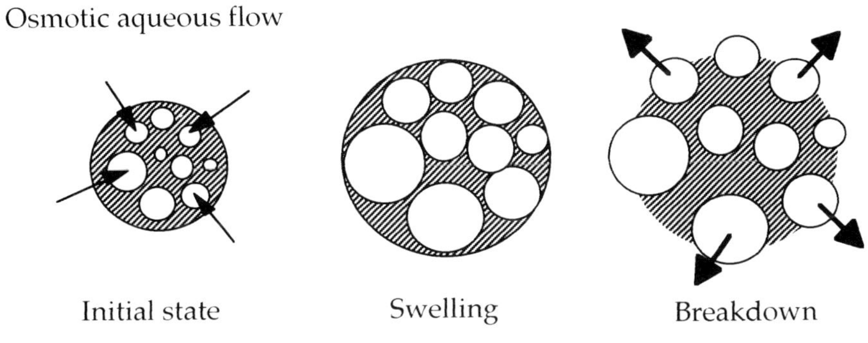

Figure 9.9 Diagram of the release mechanism by swelling–breakdown of a water-soluble molecule encapsulated in the internal aqueous phase of a multiple emulsion.

9.6 Applications

9.6.1 Pharmaceutical applications

There are as yet no pharmaceutical preparations on the market using multiple emulsions. However, these systems offer a very interesting form for drug delivery, with several potential pharmaceutical applications, and numerous *in vivo* studies have been carried out in the last 20 years. In fact, in recent years, and especially after 1990, several references to multiple emulsions as carriers for different therapeutic substances administered by various routes have been observed.

In immunology, Hearn *et al.* and Elson *et al.* were the first researchers to study, with encouraging results, this type of system for delivering antigens by an oral route to mice [12, 26].

Omotosho *et al.* carried out various studies using multiple emulsions containing 5-fluorouracil as a vehicle for anticancer agents [31, 42]. A prolonged release was obtained after intramuscular injection to rats and an important intestinal absorption found after oral administration. Lin *et al.*, with doxorubicin hydrochloride, and Higashi *et al.*, with epirupicin, obtained very good results following the administration of these substances in the form of multiple emulsions to rats (doxorubicin hydrochloride) and men (epirupicin) [22, 23]. These systems have shown a potential for drug targeting and drug delivery in cancer chemotherapy, particularly as lymphotropic agents or chemioembolisants.

Shively and Thompson showed that multiple emulsions permit the gastrointestinal absorption of vancomycin in rats, demonstrating the use of multiple emulsions as a vehicle for antibiotics [43]. Khopade *et al.*, found prolonged plasma and brain drug levels after oral and nasal administration to rats of a multiple emulsion containing rifampicin [44]. In subsequent work, the same authors obtained drug-targeting in the lungs after intravenous injection to rats of a multiple emulsion containing rifampicin [45].

Matsuzawa *et al.* showed a significant hypoglycemic effect after loop enteral administration of a multiple emulsion of insulin in the ileum and colon of rats [10]. Silva-Cunha *et al.* showed a reduction in glycemia levels after single oral administration of multiple emulsions of insulin to diabetic rats and an important change of metabolism of the diabetic rats during short-term treatment with these emulsions (Figures 9.10 and 9.11).

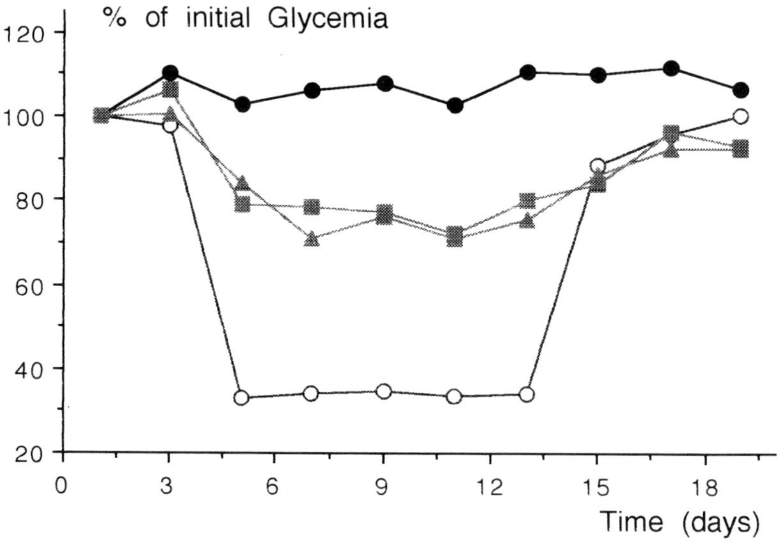

Figure 9.10 Glycemia of diabetic rats during short-term treatment. Treatment groups: insulin solution (group 1 (●) oral route and group 2 (○) subcutaneous injection), group 3 (▲) EHS and group 4 (■) ETCM (mean ±15.0% standard deviation, $n=3$) (EHS = w/o/w multiple emulsion of the soybean oil, ETCM = w/o/w multiple emulsion of the medium-chain triglycerides).

For anti-inflammatories and analgesics, Misha and Pandit showed a prolonged drug release after oral administration to mice of multiple emulsions containing pentazocine [46]. Roy and Gupta showed prolonged release of indomethacin, included in the internal aqueous phase of a w/o/w multiple emulsion, after oral administration to rabbits [6]. Finally Nakhare and Vyas obtained a prolonged release of sodium diclofenac from w/o/w multiple emulsions after intramuscular injection to rats [11].

9.6.2 Cosmetic applications

Multiple emulsions containing cosmetic active ingredients have been commer-

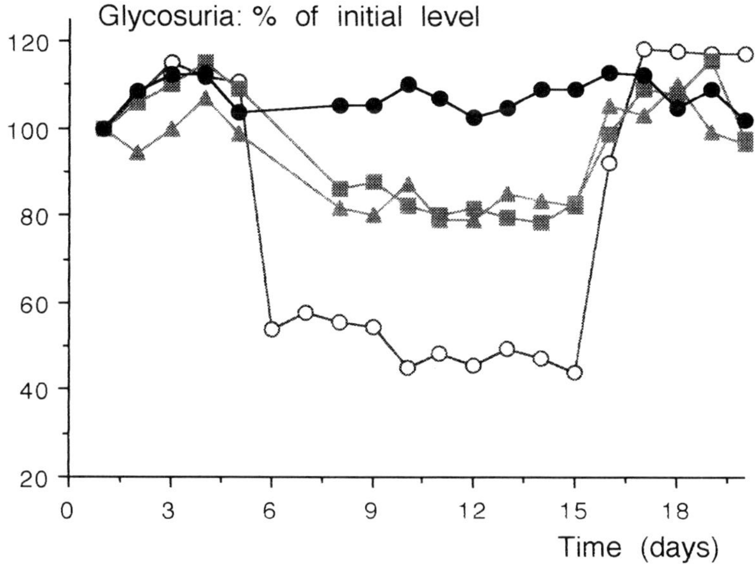

Figure 9.11 Average glycosuria for diabetic rats during short-term treatment. Treatment groups: insulin solution (group 1 (●) oral route and group 2 (○) subcutaneous injection), group 3 (▲) EHS and group 4 (■) ETCM (mean ± 9.9% standard deviation, $n = 3$) (EHS, ETCM as before).

cialised for several years and a copious patent literature exists. In essence these systems are focused on the prolonged and controlled release of active substances included in the multiple emulsion droplets. The main active substances encapsulated are hydrating agents, anti-ageing agents and perfumes. Among the most important studies published are those carried out by Ferreira *et al.* [47–49]. These authors showed that the release pattern of the same solutes (metronidazole and glucose) from w/o/w multiple emulsions is intermediary between that from a w/o simple emulsion and that from a o/w simple emulsion (the simple emulsions and the multiple emulsion being exactly of the same composition to avoid any influence of the formulation and to study only the effect of the emulsion type). Figure 9.12 shows the release of glucose (a substance with a light moisturising effect) across the skin of hairless rats. It should be noted that the release from the multiple emulsion is more rapid than from the simple w/o emulsion but slower than from the simple o/w emulsion.

It can be envisaged that w/o/w multiple emulsions could be used to obtain a comparable release of the encapsulated substances to that obtained with w/o emulsions but with the more pleasant texture of an aqueous system along with a slight moisturising effect. In fact, w/o/w multiple emulsions containing different substances (urea, sodium lactate, hyaluronic acid, vitamin C, etc.) in the internal phase have been formulated and have demonstrated very good anti-ageing and moisturising properties [34, 50].

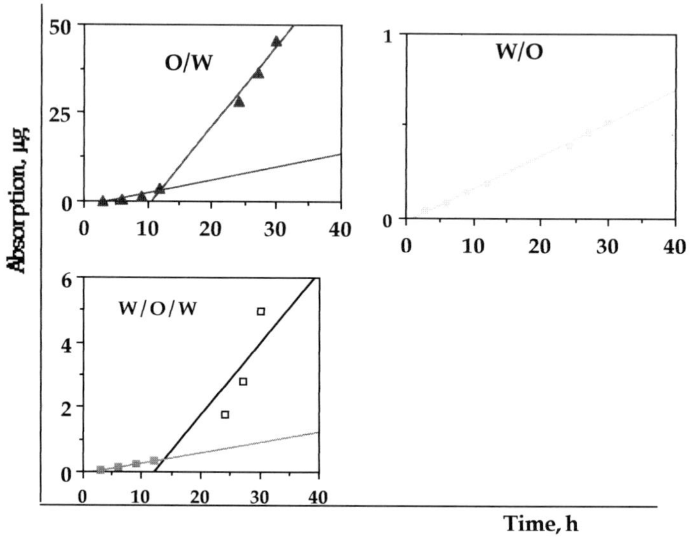

Figure 9.12 Percutaneous absorption profiles of glucose from w/o/w, o/w and w/o emulsions through hairless rat skin: typical plots of the cumulative amount of glucose as a function of time. (Values are the means, $n = 6$).

9.6.3 Applications in the food industry

The applications of multiple emulsions in food are still rather limited. However, two patents have been taken out for the following:

- w/o/w multiple emulsions for mayonnaise and a low calorie butter
- o/w/o multiple emulsion for a salad dressing (with vitamin E in the internal oily phase) and for a low-calorie margarine (whose traditional form is a simple w/o emulsion).

9.6.4 Applications in chemical processes (separation and extraction)

In these applications, which concern mainly the refining of oil and the purification of waste water, the substances in contact with the external phase should be transported passively, with the assistance of a carrier, to the internal phase. For example, Figure 9.13 illustrates the principle of the extraction and recovery of copper from the internal aqueous phase of a w/o/w multiple emulsion [51]. In the external aqueous phase, the copper is in the ionised state Cu^{2+} and as such it cannot cross the oil membrane. However, a neutral/low polarity complex formed on the external interface will be able to cross the oil membrane. The complex can be obtained by, for example, a chemical reaction between copper ions and benzoylacetone. After passing into the internal phase the complex dissociates to restore copper ions. There are also examples of extraction of phenol, acetic acid, citric acid and amines, as well as the separation of hydrocarbons by this type of process.

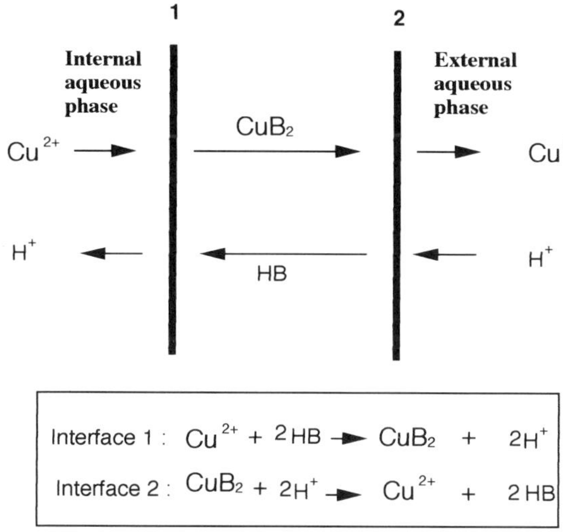

Figure 9.13 Schematic model of the extraction and recovery of copper from a w/o/w multiple emulsion.

9.7 Conclusion

After the accidental discovery of the peculiar properties of multiple emulsions these are now formulated for focused and specific applications. They are the subject of industrial research and scale-up, fulfilling the comprehensive laboratory optimisation that has taken place over the last decade.

There are indications that multiple emulsions are very interesting tools for the delivery of active ingredients because they have the same advantages as other related vesicle systems (micro- and nano-particles, liposomes, etc.) as well as those of the simpler, two-phase emulsions from which they are derived.

As a consequence of the continuous introduction to the market of new active materials and of surface-active agents of ever increasing performance and improved eco-toxicological properties, there is little doubt that these systems will attract a growing interest from researchers and industrialists in many and diverse areas.

In basic research multiple emulsions will be excellent models for the study of the biological behaviour of cells. In applied research they will be the essential tool for producing mono-particles. However, the real value of multiple emulsions will be exploited in the domain of pharmaceutical and cosmetic preparations. By analogy with liposomes it is expected that the major cosmetic manufacturers will capitalise on this technology as it allows them to encapsulate molecules of various kinds and improve their stability or effectiveness, while at the same time offering an easy application and/or a pleasant feeling on the skin.

Although w/o/w emulsions are the most widely studied, it is expected that similar considerations and perspectives will be available for the o/w/o systems, which are as yet at an early stage of exploitation. The emphasis here will be on liposoluble active materials of limited stability, which are becoming more and more common.

In the pharmaceutical field the indications from animal testing suggest that multiple emulsions could soon be tested on humans and hence confirm their potential value. They will be particularly attractive for non-traumatic administration, e.g. oral ingestions or transdermal application. The oral administration of anti-cancer drugs, of insulin, antibiotics and antigen actives is envisaged, as well as the transdermal application of anti-inflamatory drugs or steroid hormones.

References

[1] Grossiord, J.L. and Seiller, M. (eds) (1997) *Multiple Emulsions: Structure, Properties and Applications*, Editions de Santé, Paris.

[2] Garti, N., Aserin, A. and Cohen, Y. (1994) Mechanistic considerations on the release of electrolytes from multiple emulsions stabilized by BSA and nonionic surfactants. *Journal of Controlled Release*, **29** 41-45.

[3] Omotosho, J.A. (1990) The effect of acacia, gelatin and polyvinylpyrrolidone on chloroquine transport from multiple w/o/w emulsions. *International Journal of Pharmaceutics*, **62** 81-84.

[4] Oba, N., Sugimura, H., Umehara, Y., Yoshida, M., Kimura, T. and Yamaguchi, T. (1992) Evaluation of an oleic acid water-in-oil-in-water-type multiple emulsion as potential drug carrier via the enteral route. *Lipids*, **27** 701-705.

[5] Miyakawa, T., Zhang, W., Uchida, T., Kim, N.-S. and Goto, S. (1993) *In vivo* release of water-soluble drugs from stabilized water-in-oil-in-water (w/o/w) type emulsions following intravenous administrations using rats. *Biological and Pharmaceutical Bulletin*, **16** 268-72.

[6] Roy, S. and Gupta, B.K. (1993) *In vitro–in vivo* correlation of indomethacin release from prolonged release w/o/w multiple emulsion system. *Drug Development and Industrial Pharmacy*, **19** 1965-80.

[7] Terrisse, I., Seiller, M., Rabaron, A., Grossiord, J.L., Magnet, A. and Le Hen-Ferrenbach, C. (1993) Rheology: how to characterize and to predict the evolution of w/o/w multiple emulsions. *International Journal of Cosmetic Science*, **15** 53-62.

[8] Vasari, A. and Warburton, B. (1995) Improved stability of w/o/w multiple emulsions by addition of hydrophilic colloid components in the aqueous phases. *Journal of Microencapsulation*, **12** 1-5.

[9] Nakhare, S. and Vyas, S.P. (1995) Prolonged release of rifampicin from multiple w/o/w emulsion systems. *Journal of Microencapsulation*, **12** 409-15.

[10] Matsuzawa, A., Morishita, M., Takayama, K. and Nagai, T. (1995) Absorption of insulin using water-in-oil-in-water emulsion from an enteral loop in rats. *Biological and Pharmaceutical Bulletin*, **18** 1718-23.

[11] Nakhare, S. and Vyas, S.P. (1996) Preparation and characterization of multiple emulsion-based systems for controlled diclofenac sodium release. *Journal of Microencapsulation*, **13** 281-92.

[12] Elson, C.O., Tomasi, M., Dertzbaugh, M.T., Thaggard, G., Hunter, R. and Weaver, C. (1996) Oral-antigen delivery by way of a multiple emulsion system enhances oral tolerance. *Annals of the New York Academy of Science*, **778** 156-62.

[13] Okochi, H. and Nakano, M. (1996) Basic studies on formulation, method of preparation and characterization of water-in-oil-in-water type multiple emulsions containing vancomycin. *Chemical and Pharmaceutical Bulletin*, **44** 180-86.

[14] Tokgoz, N.S. (1996) Optimisation d'une emulsion multiple H/L/H renfermant de l'acide ascorbique, étude de libération et évaluation cosmétique, Thèse de Doctorat de l'Université Paris XI, Châtenay-Malabry.

[15] Silva-Cunha, A., Grossiord, J.L., Puisieux, F. and Seiller, M. (1997) Insulin in w/o/w multiple emulsions: preparation, characterization and determination of stability towards proteases *in vitro*. *Journal of Microencapsulation*, **14** 319-21.

[16] Silva-Cunha, A., Grossiord, J.L., Puisieux, F. and Seiller, M. (1997) Insulin in w/o/w multiple emulsions: biological activity after administration in normal and diabetic rats. *Journal of Microencapsulation*, **14** 321-33.

[17] Singh, S., Singh, R. and Vyas, P. (1995) Multiple emulsion-based systems carrying insulin: development and characterization. *Journal of Microencapsulation*, **12** 609-15.

[18] Dhams, G.H. and Tagawa, M. (1996) Novel multiple-phase emulsions for stable incorporation of vitamin C derivatives and enzymes, in *Proceedings of the 19th IFSCC Congress, 2, Paper N°21*, IFSCC, Sydney, pp. 79-90.

[19] Florence, A.T. and Whitehill, D. (1982) The formulation and stability of multiple emulsions. *International Journal of Pharmaceutics*, **11** 277-308.

[20] Mishra, B. and Pandit, J.K. (1990) Prolonged tissue levels of pentazocine from multiple w/o/w emulsions in mice. *Drug Development and Industrial Pharmacy*, **16** 1073-78.

[21] Lin, S.Y. and Lui, W.Y. (1992) Zero-order or first order release kinetics of water-in-oil-in-water (w/o/w) multiple emulsions of lipiodol dependent on the types of surfactants. *Biological and Pharmacological Bulletin*, **40** 2860-63.

[22] Lin, S.Y., Wu, W.H. and Lui, W.Y. (1992) *In vitro* release, pharmacokinetic and tissue distribution studies of doxorubicin hydrochloride (Adriamycin HCl®) encapsulated in lipiodolized w/o emulsions and w/o/w multiple emulsions. *Pharmazie*, **47** 439-43.

[23] Higashi, S., Shimizu, M., Nakashima, T., Iwata, K., Uchiyama, F., Tateno, S., Tamura, S. and Setoguchi, T. (1995) Arterial-injection chemotherapy for hepatocellular carcinoma using monodispersed poppy-seed oil microdroplets containing fine aqueous vesicles of epirubicin. *Cancer*, **75** 1245-54.

[24] Hino, T., Takeuchi, H., Niwa, T., Kitagawa, M. and Kawashima, Y. (1995) The analysis of drug release from diluted water/oil/water emulsions by a model of the rupture of oil membrane. *Journal of Pharmacy and Pharmacology*, **47** 1-7.

[25] Shimizu, M. and Nakane, Y. (1995) Encapsulation of biologically active proteins in a multiple emulsion. *Bioscience, Biotechnology and Biochemistry*, **59** 492-96.

[26] Hearn, T.L., Olsen, M. and Hunter, R.L. (1996) Multiple emulsions oral vaccine vehicles for inducing immunity or tolerance. *Annals of the New York Academy of Science*, **778** 388-89.

[27] Tadros, F., Dederen, C. and Taelman, M.C. (1997) A new polymeric emulsifier. *Cosmetics and Toiletries*, **112** 75-86.

[28] Magdassi, S., Frenkel, M., Garti, N. and Kasan, R. (1984) Multiple emulsions II: HLB shift caused by emulsifier migration to external interface. *Journal of Colloid and Interface Science*, **97** 374-79.

[29] Frenkel, M., Shwartz, R. and Garti, N. (1983) Multiple emulsions. I. Stability: inversion, apparent and weighted HLB. *Journal of Colloid and Interface Science*, **94** 174-78.

[30] Jager-Lazer, N., Terrisse, I., Bruneau, F., Tokgoz, S., Ferreira, L., Clausse, D., Seiller, M. and Grossiord, J.L. (1997) In fluence of lipophilic surfactant on the release kinetics of water-soluble molecules entrapped in a w/o/w multiple emulsion. *Journal of Controlled Release*, **45** 1-13.

[31] Omotosho, J.A., Whateley, T.L. and Florence, A.T. (1989) Release of 5-fluorouracil from intramuscular w/o/w multiple emulsions. *Biopharmaceutical and Drug Disposition*, **10** 257-68.

[32] Baillet, A., Pirishi, E., Vaution, C., Grossiord, J.L., Ferrier-Baylocq, D. and Seiller, M. (1994) Emulsion multiple L/H/L: formulation et suivi de la stabilité par une méthode analytique. *International Journal of Cosmetic Science*, **16** 1-15.

[33] Davis, S.S. and Walker, I. (1983) Measurement of the yield of multiple emulsion droplets by a fluorescent tracer technique. *International Journal of Pharmaceutics*, **17** 203-13.

[34] Raynal, S., Grossiord, J.L., Seiller, M. and Clausse, A. (1993) An original topical w/o/w multiple emulsion containing several active substances: formation, characterization and study of release. *Journal of Controlled Release*, **26** 129-40.

[35] Rabaron, A., Rocha-Filho, P., Vaution, C. and Seiller, M. (1989) Etude et mise en évidence des émulsions multiples L/H/L par résonance magnétique nucléaire, in *5ème Congres Internationale de Technologie Pharmaceutique*, APGI, Paris, pp. 162-71.

[36] Rocha-Filho, P., Vaution, C. and Seiller, M. (1989) Les émulsions multiples L/H/L. *STP Pharma*, **5** 652-60.

[37] Matsumoto, S., Kita, Y. and Yonezawa, D. (1976) An attempt at preparing water-in-oil-in-water multiple emulsions. *Journal of Colloid and Interface Science*, **57** 353-61.

[38] Vigie, L. (1992) Emulsions multiples H/L/H: nouveaux procédés de fabrication et de caractérisation, Mémoire pour le diplôme d'études approfondies de Biologie Cutanée et Cosmétologie, 1991–1992, Unité de formation et recherche de l'Université Paris XI, Châteney-Malabry.

[39] Kavaliunas, D.R. and Frank, S.G. (1978) Liquid crystal stabilization of multiple emulsions. *Journal of Colloid and Interface Science*, **66** 586-88.

[40] Pilman, E., Larsson, K. and Tornberg, E. (1980) Inverse micellar phases in ternary systems of polar lipids/fat/water and protein emulsification of such phases to w/o/w-microemulsion-emulsions. *Journal of Dispersion Science and Technology*, **1** 267-81.

[41] Makino, H., Fukui, H. and Matsumoto, S. (1984), in *Proceedings of the 37th Symposium on Colloid and Interface Chemistry*, Morioko, Japan.

[42] Omotosho, J.A., Florence, A.T. and Whateley, T.L. (1990) Absorption and lymphatic uptake of 5-fluorouracil in the rat following oral administration of w/o/w multiple emulsions. *International Journal of Pharmaceutics*, **61** 51-56.

[43] Shively, M.L. and Thompson, D.C. (1995) Oral biovailability of vancomycin solid-state emulsions. *International Journal of Pharmaceutics*, **117** 119-22.

[44] Khopade, A.J., Mahadik, K.R. and Jain, N.K. (1996) Enhanced brain uptake of rifampicine from w/o/w emulsions via nasal route. *Indian Journal of Pharmacy Science*, **58** 83-85.

[45] Khopade, A.J., Mahadik, K.R. and Jain, N.K. (1996) Targeting of multiple emulsions to the lungs. *Pharmazie*, **51** 558-62.

[46] Mishra, B. and Pandit, J.K. (1990) Multiple water-oil-water emulsions as prolonged release formulations of pentazocine. *Journal of Controlled Release*, **14** 53-60.

[47] Ferreira, L.A.M., Seiller, M., Grossiord, J.L., Marty, J.P. and Wepierre, J. (1994) Vehicle influence on *in vitro* release of metronidazole: role of w/o/w multiple emulsion. *International Journal of Pharmaceutics*, **109** 251-59.

[48] Ferreira, L.A.M., Seiller, M., Grossiord, J.L., Marty, J.P. and Wepierre, J. (1995) Vehicle influence on *in vitro* release of glucose: w/o, w/o/w and o/w systems compared. *Journal of Controlled Release*, **33** 349-56.

[49] Ferreira, L.A.M. (1994) Emulsions multiples H/L/H and simples L/H et H/L: étude comparative pour l'approche de la voie topique, Thèse de Doctorat de l'Université Paris XI, Châtenay-Malabry.

[50] Yazan, K., Aralp, U., Seiller, M. and Grossiord, J.L. (1995) PVP in multiple emulsions. *Cosmetics and Toiletries*, **110** 53-57.

[51] Kondo, K., Kita, K., Koida, K., Irie, J. and Nakashio, F (1979) Extraction of copper with liquid surfactant membranes containing benzoylacetone, *Journal of Chemical Engineering of Japan*, **12** 203-209.

Index